Great Explorations in Ma
(GEMS)

Lawrence Hall of Science,
University of California, Berkeley

SPACE SCIENCE SEQUENCE FOR GRADES 3–5

Unit 1 How Big and How Far?

The Space Science Sequence is a collaboration between the
Great Explorations in Math and Science (GEMS) Program
of the Lawrence Hall of Science,
University of California at Berkeley and the
NASA Sun–Earth Connection Education Forum
NASA Kepler Mission Education and Public Outreach
NASA Origins Education Forum/Hubble Space Telescope
NASA Solar System Education Forum
NASA IBEX Mission Education and Public Outreach
Special advisors: Cary Sneider and Timothy Slater
Foreword by Andrew Fraknoi

...ics and Space Administration
...GEMS Space Science Sequence was provided by the
NASA Forums and Missions listed on the title page.

Great Explorations in Math and Science (GEMS) is an ongoing curriculum development program and growing professional development network. There are more than 70 teacher's guides and handbooks in the GEMS Series, with materials kits available from Carolina Biological. GEMS is a program of the Lawrence Hall of Science, the public science education center of the University of California at Berkeley.

Lawrence Hall of Science
University of California
Berkeley, CA 94720-5200
Director: Elizabeth K. Stage

Project Coordinator: Carolyn Willard
Lead Developers: Kevin Beals, Carolyn Willard
Development Team: Jacqueline Barber, Lauren Brodsky, John Erickson, Alan Gould, Greg Schultz
Principal Editor: Lincoln Bergman
Production Manager: Steven Dunphy
Student Readings: Kevin Beals, Ashley Chase
Assessment Development: Kristin Nagy Catz
Evaluation: Kristin Nagy Catz, Ann Barter
Technology Development: Alana Chan, Nicole Medina, Glenn Motowidlak, Darrell Porcello, Roger Vang

Cover Design: Sherry McAdams, Carolina Biological Supply Co.
Internal Design: Lisa Klofkorn, Carol Bevilaqua, Sarah Kessler
Illustrations: Lisa Haderlie Baker
Copy Editor: Kathy Kaiser

This book is part of the *GEMS Space Science Sequence for Grades 3–5*.
The sequence is printed in five volumes with the following titles:
Introduction, Science Background, Assessment Scoring Guides
Unit 1: *How Big and How Far?*
Unit 2: *Earth's Shape and Gravity*
Unit 3: *How Does the Earth Move?*
Unit 4: *Moon Phases and Eclipses*

Published by Carolina Biological Supply Company. 2700 York Road, Burlington, NC 27215.
Call toll-free 1-800-334-5551.

Printed on recycled paper with soy-based inks.

ISBN 978-0-89278-334-2

UNIT 1

HOW BIG AND HOW FAR?

UNIT 1

How Big and How Far?

SESSION SUMMARIES (9 Sessions)

1.1 Thinking About Space

The unit begins with a short questionnaire to assess students' understanding of the relative sizes of the Earth, Moon, and Sun, the distances among them, and how the apparent size of an object depends on its distance from the observer. Students then read about the first "sky explorers": a sheep, duck, and chicken that were sent up in a hot air balloon. They learn the definition of *evidence,* and discuss how this expedition provided evidence that there is enough air at a certain altitude for animals (and people) to breathe. Students learn that scientific knowledge is based on evidence, and that they will be acting as scientists in the coming sessions: discussing ideas and explanations, and evaluating them based on evidence.

1.2 What's in the Sky?

Students brainstorm a list of objects in the sky, and then make drawings of objects in the sky or in space. The session concludes with an introduction to *models.* The drawings of the space objects they made are defined as a kind of model. The students learn that although models are accurate in some ways, they always contain inaccuracies, because they are not the real thing. They learn that both two-dimensional and three-dimensional models can be useful to scientists, and that space scientists find models useful because the objects they study are usually so big and far away.

1.3 Measuring Sizes of Objects

The session begins with an introduction (or review) of how to measure using metric units. Students measure a string representing the wingspan of one of the biggest birds (an albatross). In teams of two, students then measure the wingspan of the smallest bird (a hummingbird), and the lengths of four satellites.

1.4 How Big Are the Earth,Moon, and Sun?

Students are introduced to *scale models.* They predict the relative sizes of the Earth, Moon, and Sun, and then measure the diameters of scale models of these objects. They also measure how many model "Earths" fit across the diameter of a model Sun. In a class discussion, the class concludes that the Moon is very big, Earth is huge, and the Sun is comparatively super huge. A look at 3-D scale models gives them a sense of the stunning differences in size among these objects. They also learn that the Sun is a star, and is average-sized for a star.

1.5 Sizes Near and Far

In this session, students learn that objects in the sky may appear drastically different in size than they actually are, because of their different distances from Earth. The class measures a piece of paper and a student from up close and from across the room, then discusses why the measurements are different. The students are challenged to rank various objects according to actual size. Students are then given the real measurements of the objects, and have the opportunity to reorder some of the objects based on this evidence. The session concludes with a short writing assignment which helps students synthesize the concepts introduced in this session, and builds their skills in making evidence-based explanations.

1.6 Ranking Space Objects by Size

Although the primary focus of Unit 1 is on the Earth, Moon, and Sun, this session temporarily digresses to explore other space objects that your students may be curious about. The focus is still on size, allowing students to put space objects into context based on the sizes of the Earth, Moon, and Sun. They learn that there are many space objects out there that are inconceivably big. The session also helps deepen their understanding that huge space objects may look small because they are far away.

Each team categorizes a set of sky object cards by size, comparing each sky object to the size of a school, the Earth, the Moon and the Sun. When they have completed this task, the teacher leads the class in a *Tour of Sky Objects*. Information about the sizes of the objects, as determined by space scientists, is revealed, and students adjust their sorts of the cards accordingly. The tour serves the dual purpose of revealing the actual relative sizes of the objects and providing an opportunity for students to gaze in wonderment at beautiful and mysterious space images.

1.7 How Far Away Are They?

Using a new scale ruler, student pairs circulate to learning stations around the classroom to measure the distance between the ground and 13 different objects—including the tallest mountain, a cloud, a hot air balloon, an airplane, satellites, the International Space Station, and the Moon. They make their measurements on scale drawings. The students discuss their findings. The relative distances of all objects are discussed, including how much farther away the Moon is than any of the other objects measured. They also discuss which objects are in Earth's atmosphere and which are in space.

1.8 Comparing Distances

Students begin by reading *Jumping from the Edge of Space*, which is the story of the first person to skydive back to Earth from "the edge of space." The class puts the altitude of the skydive in context of the sky objects they measured in Session 1.7. Using a different scale, the class compares the relative distance to the Moon with the other distances they have measured. Students go outside (or into a long hallway or large room) to pace off the distance to the Sun, and contrast this distance with the distance to the Moon. Back in the classroom, students discuss a question about the relative distances of the Moon and Sun in small groups called *evidence circles*. They apply their new evidence as they discuss the question.

1.9 How Our Scale Ideas Have Changed

Students learn that either through magnification or through getting closer with a spacecraft, or a combination of the two, we can see larger images of sky objects. They learn that telescopes are very useful tools for space scientists.

A short demonstration reviews relative sizes of sky objects and solidifies students' understanding of how apparent size depends on distance from the viewer. The teacher holds up a dot representing the size of Venus, and students compare it with a "Planet X" dot at arm's length. The teacher backs up and the students note when the dot representing Venus is far enough away that it appears to be the same size as the dot representing Planet X. They discuss how far away a dot representing Jupiter and another representing the Sun would need to be to appear as small as they do in the sky. They also review that although our Sun is an average-sized star, it looks much bigger than other stars because it is much closer to us.

In evidence circles of four, the students are given three inaccurate models of the Solar System. They apply the evidence that they have gathered about scale as they evaluate a model for accuracy in terms of size and distance. Their discussion is continued with the whole class. To conclude the unit, the students take the *Post–Unit 1 Questionnaire* on size and distance to find out how their ideas have changed.

SESSION 1.1

Thinking About Space

*Finding out what students already understand (and misunderstand) is helpful to the teacher in gauging instruction. Helping students articulate what they think is also an important step in enabling them to move beyond their initial understandings.

Unit Goals

Size: Some sky objects are relatively small, and some are huge.

Distance: Some objects are relatively close to Earth and some are very far away.

Distance of sky objects from us affects their apparent size: Large objects appear small when far away.

Overview

The unit begins with a short questionnaire to assess students' understanding of the relative sizes of the Sun, Earth, and Moon, the distances between them, and how the apparent size of an object depends on its distance from the observer. A similar post-unit questionnaire will be administered in Session 1.9 to assess how their ideas about these concepts have changed.

Students then discuss what the first exploration of the sky might have been like, and read about the first "sky explorers," a sheep, a duck, and a chicken that were sent up in a hot air balloon. They learn the definition of scientific evidence, and discuss how this expedition provided evidence that there was enough air at a certain altitude for animals (and people) to breathe.

Students learn that scientific knowledge is based on evidence and that they will be acting as scientists in the coming sessions, discussing ideas and explanations, and evaluating them based on evidence. Six key concepts are posted on what will become the class "concept wall" for the whole unit.

Thinking about Space	Estimated Time
Introducing the unit and the practice questionnaire	15 minutes
Taking the *Pre-Unit 1 Questionnaire*	15 minutes
Reading: *The Adventures of a Sheep, Duck, & Chicken*	10 minutes
Discussing the reading	10 minutes
Introducing key concepts	10 minutes
TOTAL	**60 minutes**

What You Need

For the class

- ❑ overhead projector or computer with large screen monitor/LCD projector
- ❑ overhead transparencies of the two pages of the *Practice Questionnaire*, from the transparency packet or CD-ROM file
- ❑ overhead marker
- ❑ sentence strips to record 6 key concepts
- ❑ wide-tip felt pen
- ❑ *optional:* 1 meter stick or metric measuring tape

TEACHER CONSIDERATIONS

SCIENCE NOTES

Nature of Science—Evidence and Scientific Explanations: The concept of evidence is crucial to an understanding of what science is, and to the process of scientific inquiry. Making explanations based on evidence is one of the central strands in the development of science inquiry abilities, as is the ability to distinguish evidence from inference.

In science, knowledge is based on evidence. As defined in the *National Science Education Standards* (NRC, 1996), "evidence consists of observations and data on which to base scientific explanations." The Standards suggest that students in primary grades develop basic understandings of science inquiry, including that, "Scientists develop explanations using observations (evidence) and what they already know about the world (scientific knowledge). Good explanations are based on evidence from investigations."

Space scientists study remote objects, so it's not common for them to have a physical specimen to show as evidence. The evidence they gather is frequently composed of careful observations, images, measurements, and other data that can be verified by others. A scientist's findings must be critically tested, examined, and verified by other scientists before being considered valid scientific evidence. These ideas will be reinforced across the entire sequence.

Throughout Unit 1, students will have opportunities to behave as scientists—to gather evidence about the size, distance, and movements of space objects, discuss it, and try to come up with explanations that best match all the evidence. Students will learn to distinguish evidence-based explanations from non-scientific explanations, which may not be based on evidence at all, or may be based only on selective evidence. Scientists base their explanations on all the available evidence. It is not a scientific approach to look only for evidence that supports an idea, but ignore evidence that doesn't.

Key Vocabulary

Science and Inquiry Vocabulary

Evidence
Scientific Explanation
Model
Scale Model
Prediction
Scientist
Three Dimensional (3-D)
Two Dimensional (2-D)

Space Science Vocabulary

Atmosphere
Satellite
Orbit
Diameter
Sphere
System

For each student:

- ❑ 1 copy of the *Pre–Unit 1 Questionnaire* (two pages), from the student sheet packet
- ❑ 1 copy of the reading, *The Adventures of a Sheep, a Duck, and a Chicken*, from the student sheet packet
- ❑ *optional:* a copy of page 2 of the reading, from the student sheet packet

Getting Ready

1. Arrange for the appropriate projector format (computer with large screen monitor, LCD projector, or overhead projector) to display images to the class.

2. If you will not be using the CD–ROM, make two overhead transparencies, one of each of the pages of the *Practice Questionnaire*.

3. Make a copy for each student of the *Pre–Unit 1 Questionnaire.*

4. Make a copy for each student of the reading *The Adventures of a Sheep, a Duck, and a Chicken.* This is a one-page reading with an optional second page at a slightly higher reading level. Everyone will read the first page as part of the core of the session. Depending on your students' reading abilities and the time available, decide whether you will copy and have them read page 2.

**See page 221 to see how your concept wall might look at the end of Unit 1, with all key concepts posted.*

5. Choose a wall or bulletin board where you can post key concepts on sentence strips (or chart paper). This will be your "concept wall," to which you will add a total of 30 key concepts for Unit 1. Although substantial space is required, making space for the concept wall is strongly recommended. Making learning explicit to students, and constructing a growing representation of knowledge on the wall, are powerful learning devices and essential components of this unit. We suggest you arrange the sentence strips under two headings: 1) *What We Have Learned About What Scientists Do* and 2) *What We Have Learned About Space Science.*

6. Use a map or your car's odometer to find a landmark about 2 kilometers (about 1.25 miles) from school to help students understand how high the hot air balloon traveled. Any store, restaurant, or other landmark known to your students will do.

Unit Goals

Size: Some sky objects are relatively small, and some are huge.

Distance: Some objects are relatively close to Earth and some are very far away.

Distance of sky objects from us affects their apparent size: Large objects appear small when far away.

TEACHER CONSIDERATIONS

TEACHING NOTES

Reading level: The reading level of page 1 of the reading is appropriate for most third and fourth graders. Page 2 is for students who are interested in further information on the topic, and who are able to read at a slightly higher level.

Units of Measure: Metric units of measure are used throughout the *Space Science Sequence*. In the student reading in Session 1.1, the hot air balloon is described as being about 13 meters in diameter. If your students are not already familiar with meters, you might want to show them a meter stick or tape and have them estimate the size of the balloon.

Key Vocabulary

Science and Inquiry Vocabulary

Evidence

Scientific Explanation

Model

Scale Model

Prediction

Scientist

Three–Dimensional (3-D)

Two–Dimensional (2-D)

Space Science Vocabulary

Atmosphere

Satellite

Orbit

Diameter

Sphere

System

7. Write the following three key concepts on sentence strips with a wide-tip felt pen, and have them ready to post during the session under the heading, *What We Have Learned About What Scientists Do.*

Evidence is information, such as measurements or observations, that is used to help explain things.
Scientists base their explanations on evidence.
Scientists question, discuss, and check one another's evidence and explanations.

8. On sentence strips, write the following three key concepts and have them ready to add to the concept wall at the end of the session. These will be the first three concepts under the heading, *What We Have Learned About Space Science.*

Earth is surrounded by an atmosphere of air.
Beyond Earth's atmosphere is what we call space.
People have been wondering and learning about space for a long time.

Introducing the Unit

1. Introduce the unit on space science. Tell students they will get to be space scientists in the coming weeks. They will learn about the sizes and distances of objects in space.

2. Introduce purpose of questionnaire. Tell your students you have a questionnaire for them to take, which will help you find out what they already know about space.

3. Explain that they will probably not know all the answers. Say that it's fine to think about it and make their best guess. Say that the questionnaire will not be used for a grade. They will take the questionnaire again in a couple of weeks to find out how much they have learned.

Unit Goals

Size: Some sky objects are relatively small, and some are huge.

Distance: Some objects are relatively close to Earth and some are very far away.

Distance of sky objects from us affects their apparent size: Large objects appear small when far away.

TEACHER CONSIDERATIONS

TEACHING NOTES

Time Management: Your students will probably be excited about talking and asking questions about space. Although there isn't time for much discussion during this initial class session, assure students that they will have opportunities later in the unit to talk about space.

What One Teacher Said

"One thing I have noticed over the years is that any time students have the opportunity to talk about space science, they do."

Key Vocabulary

Science and Inquiry Vocabulary

Evidence

Scientific Explanation

Model

Scale Model

Prediction

Scientist

Three–Dimensional (3-D)

Two–Dimensional (2-D)

Space Science Vocabulary

Atmosphere

Satellite

Orbit

Diameter

Sphere

System

PRACTICE QUESTIONNAIRE, PAGE 1

Session 1.1 Transparency
Practice Questionnaire, Page 1

1. One of the pictures below shows the correct sizes of an orange, a pumpkin, and a cherry compared with one another. Which is the best? Circle the letter of the best one.

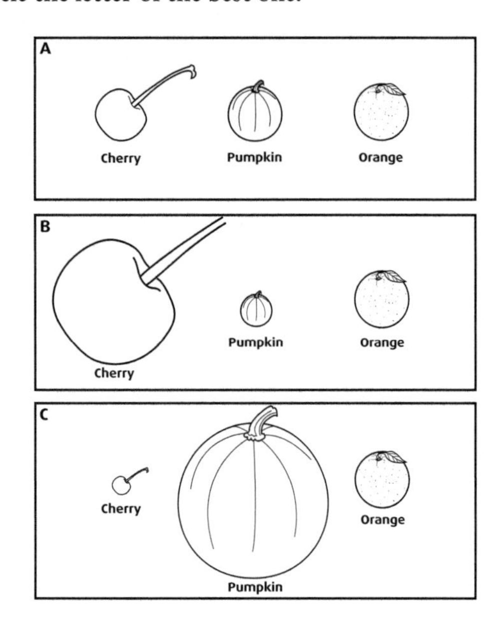

2. Why does ice cream melt on a sunny day? Circle the letter for the best answer:

A. The ice cream melts because it is cold outside.
B. The ice cream melts because it is hot outside.
C. The ice cream melts because it is green.
D. The ice cream melts because it is daytime.

The Practice Questionnaire

1. Fill out the *Practice Questionnaire* together. Say that before they fill out the real questionnaire, you would like the class to discuss some practice questions together. The practice questions are not about space, but will get them used to the kind of questions they will answer on the real questionnaire.

2. Question #1 on the practice questionnaire. Show page 1 of the *Practice Questionnaire*. Have someone read the question: *One of the pictures below shows the correct sizes of an orange, a pumpkin, and a cherry compared with one another. Which is the best? Circle the letter of the best one.* Before you elicit the correct answer [C], make sure all students are clear on the point of the question and understand how to answer.

- The boxes labeled A, B, and C are the options. Ask, "Do we circle a whole box or picture?" [No, just circle the letter of the best answer.]
- Why might someone pick B? Agree that almost everyone knows a pumpkin is really bigger than a cherry. But someone might pick B because a cherry could *look* bigger than a pumpkin if the cherry were up close and the pumpkin were far away. Reread the question and emphasize that it is asking how big they think these objects *really are*, compared with one another, not how big or small they might *look*.

3. Question #2. Choose the Best Written Answer. Ask, "Could D be correct?" [Yes, because sunny days can be hot.] "Is it the *best* answer?" [B is probably better, because heat is more directly the cause of ice cream melting than daytime.] "Should you circle the whole sentence?" [No, circle only the letter B.]

TEACHING NOTES

Adjusting for Student Experience: If your students are experienced with the kinds of questions on the Unit 1 questionnaire, you may decide to skip the *Practice Questionnaire*, or go over only selected questions together.

Make Sure Students Understand the Questions: A question may be answered incorrectly because students don't understand what the question is asking or what the task is. In order to get a more accurate assessment of *what students know*, it's worth the extra effort to ensure that they understand the procedures in each question. Make it clear that if they are unsure about what they are supposed to do as they are taking the questionnaire, students should ask for assistance.

PRACTICE QUESTIONNAIRE, PAGE 2

Session 1.1 Transparency
Practice Questionnaire, Page 2

3. The person in the picture just kicked the ball. Where will the ball go? Draw an arrow to show where you think the ball will go.

4. The answers below compare the distances among a person's feet, knees, and head. Which is the best? Circle the letter of the best answer.

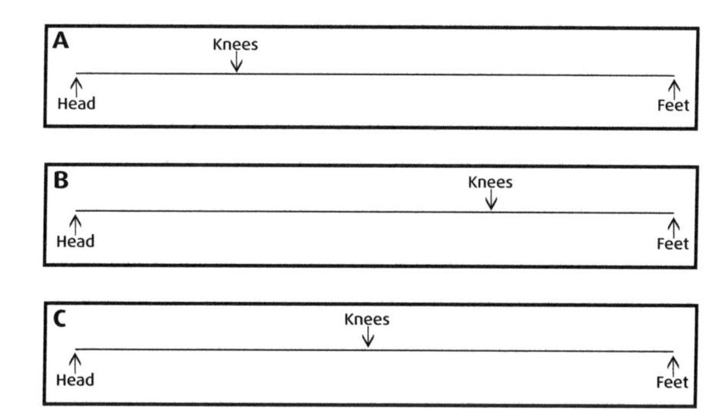

5. What is happening in the picture? Explain.

PRE-UNIT 1 QUESTIONNAIRE, PAGE 1

Name:________________________

Pre-unit 1 Questionnaire

1. One of the pictures below shows the correct sizes of the Sun, Earth, and Moon compared with one another. This question is about the real sizes, not how big they look in the sky. Which is the best? Circle the letter of the best one.

A: Sun, Moon, Earth

B: Sun, Moon, Earth

C: Sun, Moon, Earth

D: Sun, Moon, Earth

E: Sun, Moon, Earth

4. Question #3: Draw Your Best Answer. Ask a student first to describe in words where the pictured ball will go, and then ask how you could draw that path. [Show an arrow going away from the foot and up, or bouncing—or whatever the student predicted.]

5. Question #4: Compare Distances. Ask, "Are people's knees closer to their heads or their feet?" Help the class understand that each line on the drawings represents a person lying down, with the head on the left and feet on the right. Each picture shows the knees at a different distance from their head and feet. Say they will again choose the best one and circle the letter. [B.]

6. Question #5: Use Writing to Explain a Picture. Help students verbalize an explanation for what is happening in this picture. [For example, "It is a picture of rain falling, and a person is standing under a tree with no rain falling on him."]

Taking the Pre-Unit 1 Questionnaire

1. Tell students not to help one another. Explain that each student will now receive a copy of the questionnaire about space. Say that usually during activities it is fine to help each other, but this time you want them each to answer the questions without talking to anyone else. The questionnaire is designed to find out what each student thinks about these questions.

2. Tell them to ask for help if they don't understand what to do. Say that if they don't understand what the question is asking them to do, they should raise a hand for help.

3. Distribute questionnaires and have students begin. Give each student a copy of the *Pre–Unit 1 Questionnaire.*

4. Collect the questionnaires. When everyone is finished, collect the questionnaires, making sure that students have put their names on them.

TEACHING NOTES

Classroom Management: It's helpful to have an adult volunteer to help the first time you administer this questionnaire. Be sure adult helpers know they should help only with the procedure and comprehension of the questions. They should not give hints or help with the answers.

QUESTIONNAIRE CONNECTION

We have included *Questionnaire Connection* notes in later sessions of the unit to help you highlight when an activity or discussion relates to specific questions on the questionnaire.

Order of Questions on the Pre and Post Questionnaires: The *Post-Unit 1 Questionnaire* used in Session 1.9 is identical in content to the *Pre-Unit 1 Questionnaire,* but the placement of some questions and the order of the possible answers is different. This is to encourage students to answer thoughtfully, rather than simply to remember the order of answers.

PRE-UNIT 1 QUESTIONNAIRE, PAGE 2

Name:________________________

Pre-unit 1 Questionnaire continued

2. Why does the Sun look much bigger than the stars we see at night? Circle the letter of the best answer.
   A. The Sun is much bigger than the stars.
   B. The Sun looks bigger because it is closer to us than the stars.
   C. The Sun looks bigger because it is farther away from us than the stars.
   D. The Sun can be seen only in the daytime.

3. The answers below compare the distances among the Sun, Earth, and Moon. Which is the best? Circle the letter of the best one.

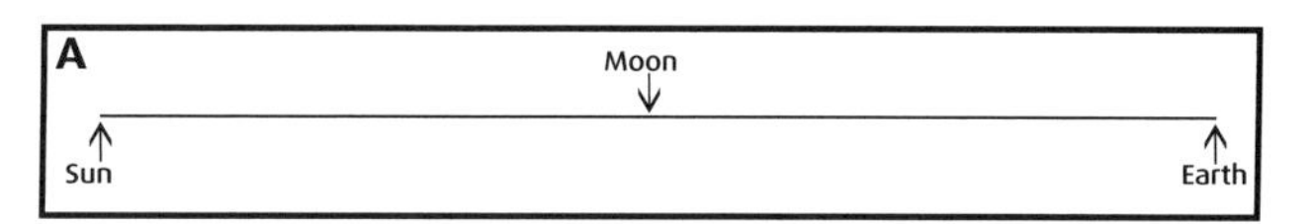

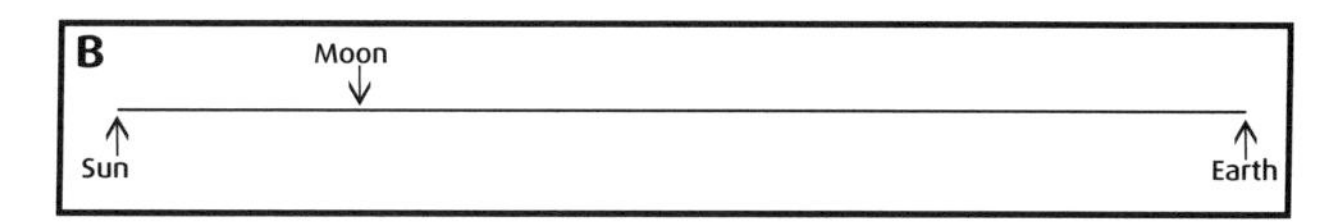

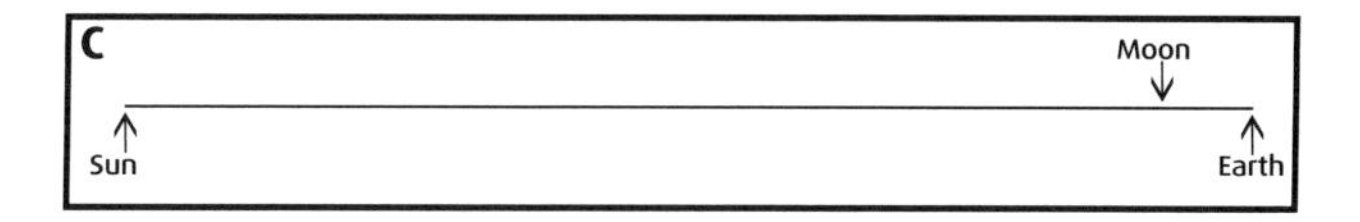

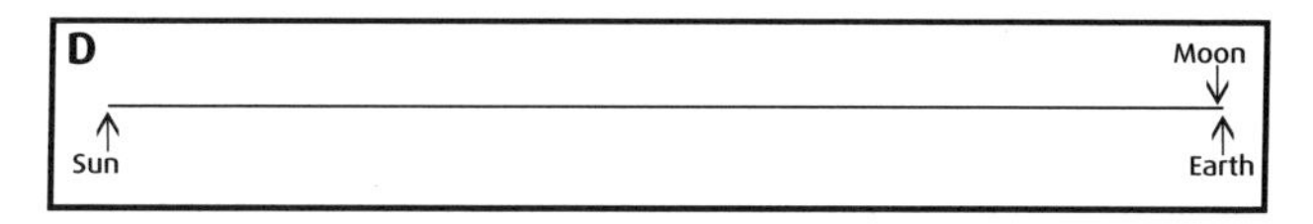

4. Why do the Sun and Moon look as though they are about the same size in the sky? In your answer, explain how big and how far away they are.

____

____

Scoring the *Pre-Unit 1 Questionnaire*: It is important to look over all of your students' questionnaires before getting too far into the unit so you can identify areas that may need more or less instruction and any major misconceptions students may have. If you wish, the questionnaire can be scored using the scoring guide on page 68.

THE ADVENTURES OF . . . PAGE 1

Name:____________________________

The Adventures of a Sheep, a Duck, and a Chicken

People have always looked at the sky and wondered what was up there. What are all those points of light? What is that bright, round thing that looks like it comes up in the morning and goes down at night? What is that bright, white, round thing you can sometimes see at night? How high in the sky are all those things? How big are they?

People also wondered if they would ever be able to fly into the sky to explore it.

In 1783 people sent the first explorers into the sky.

But it wasn't people who went on the first trip. People worried that there would be less air higher in the sky. They thought people might die because they couldn't breathe.

So instead of people, the first riders in a hot air balloon were a sheep, a duck, and a chicken. People sent these animals higher in the atmosphere to see what would happen. If the animals survived, people would have evidence that there was enough air to breathe up there.

The balloon was about 13 meters in diameter. The sheep, duck, and chicken rode the balloon high into the sky for eight minutes. They rose about 2,000 meters high. When the balloon landed, people could see that flying so high had not hurt the animals. This was evidence that there was enough air to breathe higher in the atmosphere.

A few weeks later, two Frenchmen made the next trip into the sky. Since then, many other people have traveled higher and higher into the atmosphere. They have found out that people were right to worry about not having enough air. The higher you fly, the less air there is. People who fly high enough can't breathe without air tanks. People have traveled beyond Earth's atmosphere into outer space. Some people have even traveled all the way to the Moon, where there is no air.

But remember that the ones to start all this were a sheep, a duck, and a chicken.

Introducing the Reading: *The Adventures of a Sheep, a Duck, and a Chicken*

1. Access students' prior knowledge. Ask questions to prepare your students for the reading:

- Have you heard of balloons that carry people up into the sky?

- What makes these balloons go up? [Some are filled with hot air, and some are filled with light gases, such as helium and hydrogen. Usually, the travelers sit in a basket or other carrier attached to the balloon.]

2. Define Earth's atmosphere. Ask, "What is the Earth's *atmosphere*?" [The air or gas around the Earth.] Tell them that Earth's atmosphere is made of gases, including oxygen which our bodies need to breathe. Explain that there is less and less air as you go higher in the sky, but people didn't always know that. Beyond the atmosphere is *space,* where there is no air.

3. Imagine the first balloon flight. Say, "Imagine you lived in 1783 before anyone had ever traveled into the atmosphere, and long before airplanes were invented." Tell students to imagine that a balloon was just invented that could take people up in the sky. People must have wondered: "Will there be enough air to breathe up there?"

4. Collecting information about the atmosphere and beyond. Ask, "What could be done to answer this question, without sending *people* up?" Discuss a few ideas.

5. Introduce *evidence*. Explain that scientists look for clues to help explain things or to help show that something is true. Evidence most often comes from what we gather through our senses: seeing, hearing, feeling, smelling, and tasting. These clues are called *evidence*. Say that back in 1783, scientists were looking for evidence about whether or not people could breathe if they went high up in the atmosphere.

TEACHER CONSIDERATIONS

TEACHING NOTES

Expectations for learning about the atmosphere: A complete picture of the atmosphere depends on understanding concepts about the Earth's size and spherical shape. Students will learn about Earth's scale in later sessions of Unit 1, and deepen their understanding of the Earth's shape in Unit 2. For now, whether or not students yet grasp the scale or shape of the Earth, they should understand that the Earth has an atmosphere, that it is made of a mixture of gases we call air, that there is less air as you go higher above Earth's surface, and that there is no air in space.

.

Review challenging words: Depending on the reading level of your students, you may want to review some or all of these potentially difficult words before the reading: *breathe, survive, French, explore, oxygen, atmosphere.*

Purposes of the Reading:

- Provides an historical example of how people have performed tests and experiments to find evidence to answer questions. This is one of the most important components of scientific investigation.
- Introduces the idea of space exploration — people sending aircraft, spacecraft, and themselves up into the sky on missions to explore.
- Reinforces concepts about the atmosphere: that there is air in our atmosphere, that there is less and less air as you go higher toward space, and that there is no air in space.
- Many students tend to think that gravity is caused by air. This reading helps prepare them for Unit 2, in which they learn that there is gravity on the Moon and on other planets, even when there is no air.

THE ADVENTURES OF . . . PAGE 2

Name:______________________________

The Adventures of a Sheep, a Duck, and a Chicken continued

The First People to Fly in Hot Air Balloons

The healthy chicken, duck, and sheep balloonists were evidence that there was enough air to breathe higher in the atmosphere. After this test, people were ready to try traveling into the atmosphere themselves.

On the next flight, two Frenchmen were in the basket. One was a science teacher and the other was a soldier. A rope kept the balloon from flying away.

Next, the science teacher and the soldier were ready to try a balloon flight with no rope. The king of France and many others watched. The balloonists flew 100 meters high for 25 minutes, and landed 9 kilometers away. Two years later, the science teacher died in a balloon accident while trying to fly across the English Channel.

Other Early Flight History

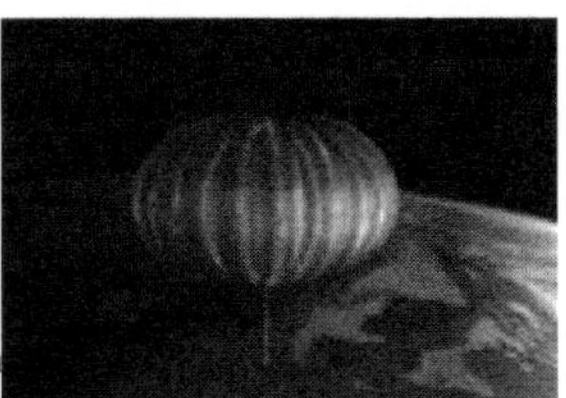

- In 1784 a thirteen-year-old boy named Edward Warren became the first person in America to fly in a balloon.
- In 1804 a scientist flew 7,016 meters high to study the air. He became dizzy, felt his heart beating faster, had trouble breathing, and finally passed out. When his balloon began sinking back down, he woke up. The problems he had were evidence that there was not enough oxygen to breathe high u in the atmosphere.
- In 1903 the Wright brothers flew the first airplane a distance of over 37 meters.
- In 1960 people developed a new, safer kind of hot air balloon, and hot air balloons became popular again.

The Injured Duck

When the balloon carrying the sheep, duck, and chicken landed, people saw that the duck did not look well. They wondered if flying so high had hurt the duck in some way. But others had seen the n left the ground. The duck's wing was broken.

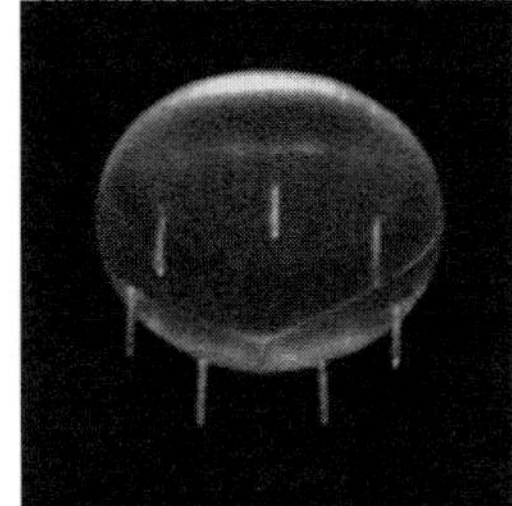

onster Attack from the Sky

ne of the first test balloons landed in a town. Some people who ived in the town thought the balloon was a monster attacking them rom the sky. They tore it to pieces with their farm tools. Luckily, o people (or animals) were riding in the balloon.

6. **Introduce the reading.** Write the following question on the board: *Can people breathe if they go high in the atmosphere?* Tell students they're going to read about the first time Earth explorers went up into the sky. Tell students that, as they read, they should look for evidence to help answer this question.

7. **Read *The Adventures of a Sheep, a Duck, and a Chicken.*** Depending on your students' reading abilities, you may want to have them do independent, paired, or shared reading.

After the Reading

1. **Discuss the Reading.** Use the following questions to prompt discussion. Be sure to emphasize the concept of evidence:

- Why do you think they sent animals instead of people up in the balloon on the first trip? [They wanted to gather evidence about whether there was air to breathe, but didn't want to risk human lives.]
- After the first balloon trip, what evidence showed that there was enough air to breathe up there? [The sheep, duck, and chicken survived.]
- How high did the balloon go? [The balloon went up 2 kilometers. Tell them that 1,000 meters = 1 kilometer. Give students an idea of how far away 2 kilometers would be on land by giving a landmark about 2 kilometers (1.25 miles) from the school.]
- Have you ever traveled to high mountains, where there is less air? What was it like?
- Have you ever traveled on an airplane? Was there enough air to breathe? Were the windows open? [When airplanes fly low, people can open windows and breathe the air. When they travel very high, the windows are usually kept closed, trapping more oxygen inside, so people can breathe.]
- Do you think there would be enough air to breathe if the sheep, duck, and chicken traveled even higher? [Note: At 5 kilometers (almost 3 miles) high and up, explorers usually use oxygen tanks to breathe. Mt. Everest is about 9 kilometers high.]

Unit Goals

Size: Some sky objects are relatively small, and some are huge.

Distance: Some objects are relatively close to Earth and some are very far away.

Distance of sky objects from us affects their apparent size: Large objects appear small when far away.

TEACHER CONSIDERATIONS

PROVIDING MORE EXPERIENCE

You might want to use some of these additional questions for discussion or as writing prompts if time allows:

- How big was the balloon? [The balloon was 13 meters in diameter.] Briefly define a meter as about one "giant step," or about 3 inches more than a yard, and say that *diameter* is the distance across the widest part of the balloon. Quickly pace out approximately 13 meters in your classroom, so students can get an idea of the diameter of the balloon.
- Can you think of any examples in modern times when people might come up with non-scientific explanations for things they do not understand?
- What other explorations of the sky have you heard of?
- What are some questions you have about what's in the sky?
- Would you ever want to travel into the sky in a balloon or a spaceship? Why or why not?

ASSESSMENT OPPORTUNITY

Critical Juncture: Throughout this unit and the rest of the sequence, students will have opportunities to practice backing up their explanations with evidence. If you think your students need more examples of how scientific explanations are supported by evidence, you might want to take some time for the activities described below.

PROVIDING MORE EXPERIENCE

1. The hurt duck example. Read the following passage aloud:

When the balloon carrying the sheep, duck, and chicken landed, people saw that the duck did not look well, and had a broken wing. They wondered if it had been hurt by the altitude. But others had seen the sheep kick the duck as the balloon left the ground.

Point out that if someone said the high altitude had hurt the duck, they were looking at only *some* of the evidence (a hurt duck and the recent balloon trip). The scientific explanation (that the duck didn't look well because the sheep had kicked it) included *all* the evidence. Good scientists base their explanations on all available evidence. They are able to change their minds when new evidence is reported.

continued on page 109

Key Vocabulary

Science and Inquiry Vocabulary

Evidence
Scientific Explanation
Scale Model
Prediction
Scientist
Three Dimensional (3-D)
Two Dimensional (2-D)

Space Science Vocabulary

Atmosphere
Satellite
Orbit
Diameter
Sphere
System

2. Define scientific explanations. Tell the class that scientists try to explain how the world works. Say that a *scientific explanation* is based on evidence. Scientists discuss and check one another's evidence and explanations.

3. Point out this example of an early attempt at gathering evidence and making explanations. Tell them that the flight of the sheep, duck, and chicken is an example of doing a test to answer a question scientifically. The test (sending up three different species of animals) provided them with some evidence—the animals lived. The explanation they came up with was based on this evidence. Their explanation was that the Earth's atmosphere has enough air for animals to breathe—at least up as far as the balloon went.

Introducing Key Concepts

1. Cast students in the role of scientists. Tell students that, like scientists, they will be gathering evidence about sky objects in the coming class sessions. They will also make their best explanations based on the evidence.

2. Introduce the concept wall. Tell students that throughout the space science unit, they will add key concepts to the class *concept wall* to show what they are learning. Tell them that key concepts are important ideas to remember. They will refer back to the wall and use it to help them remember, discuss, and explain what they have learned. It will also show how things are connected to one another.

3. Post the first three key concepts on the concept wall. Say that some of the things they have learned about scientific evidence are key concepts. Post the three sentence strips you prepared earlier under the heading, *What We Have Learned About What Scientists Do*, and remind students of how they relate to the previous discussion.

Evidence is information, such as measurements or observations, that is used to help explain things.
Scientists base their explanations on evidence.
Scientists question, discuss, and check each other's evidence and explanations.

Unit Goals

Size: Some sky objects are relatively small, and some are huge.

Distance: Some objects are relatively close to Earth and some are very far away.

Distance of sky objects from us affects their apparent size: Large objects appear small when far away.

PROVIDING MORE EXPERIENCE *CONTINUED*

2. The "balloon monster" example. Read the following passage aloud:

One of the first flying balloons tested without people in it landed in a town. Some people who lived in the town thought it was a monster attacking them from the sky. They tore it to pieces with their farm tools.

Ask if "balloon monster" is a scientific explanation for what the people observed. Tell students that this is an example of people coming up with an explanation in a non scientific way. There was no evidence that the balloon was a monster.

3. A role play about a "flying saucer." Tell students you will play the role of the person who says they saw a flying saucer from another planet. The class will play the role of scientists trying to find out if there is enough evidence to verify this claim. Explain that scientists don't agree that something is real unless they have scientific evidence. Say:

"I saw a light in the sky and it looked like it was flying away from me. I think it was a flying saucer from another planet."

Ask whether or not there could have been a different explanation for the evidence (a light in the sky flying away). Have students list some other possible explanations for the evidence. Then, have the class vote on the issue. Remind them that as scientists, they can only vote that it was a flying saucer if they think the evidence cannot be explained in any other reasonable way.

In conclusion, say that some scientists do think there may be extraterrestrials far from Earth. But although many people say they have seen extraterrestrials, no one has ever been able to give scientists solid, convincing evidence to support the claim.

4. Writing Activity. Ask students to rewrite the story in a different genre. For example, they could write it as a news article or interview, and emphasize the evidence that was gathered by scientists. For a more general creative writing assignment, they could write a postcard or letter from the perspective of a character in the story.

Key Vocabulary

Science and Inquiry Vocabulary

Evidence
Scientific Explanation
Model
Scale Model
Prediction
Scientist
Three–Dimensional (3-D)
Two–Dimensional (2-D)

Space Science Vocabulary

Atmosphere
Satellite
Orbit
Diameter
Sphere
System

4. Post the next three key concepts. Tell students they have learned three key concepts about space science. Post the sentence strips you prepared earlier on the concept wall, under the heading, *What We Have Learned About Space Science.*

Earth is surrounded by an atmosphere of air.
Beyond Earth's atmosphere is what we call space.
People have been wondering and learning about space for a long time.

Unit Goals

Size: Some sky objects are relatively small, and some are huge.

Distance: Some objects are relatively close to Earth and some are very far away.

Distance of sky objects from us affects their apparent size: Large objects appear small when far away.

TEACHER CONSIDERATIONS

OPTIONAL PROMPTS FOR WRITING OR DISCUSSION

Optional prompts are provided at the end of each session in the unit. They can offer an additional opportunity for students to process information and ideas. Prompts may be used for science journal writing during class or as homework. Or they could be used for a discussion or during a final student sharing circle in which each student gets a turn to share.

1. What are three things you have learned about space science today?
2. Design a mission to find out if a person could survive parachuting from very high in the sky.
   - What problems would you be worried about?
   - What test would you do to get the evidence?
   - Would you send living things on the first mission?
   - What evidence would tell you if it were possible to survive?

Key Vocabulary

Science and Inquiry Vocabulary

Evidence

Scientific Explanation

Model

Scale Model

Prediction

Scientist

Three–Dimensional (3-D)

Two–Dimensional (2-D)

Space Science Vocabulary

Atmosphere

Satellite

Orbit

Diameter

Sphere

System

SESSION 1.2
What's in the Sky?

Overview

Students brainstorm a list of objects that can be seen in the night and day sky. They then list space objects they have heard of, but that cannot be seen from Earth with the naked eye.

The students then make drawings of objects in the sky or in space. These can be objects from the class list, or others. These informal drawings will serve as a basis for discussions of scientific models.

The session concludes with an introduction to models. First, the drawings of space objects they made are examined and defined as a kind of model. The students learn that although models are accurate in some way(s), they always contain inaccuracies, because they are not the real thing. Students critique a simple model such as a toy car, and describe all the inaccuracies they can think of in the model. They learn that both two-dimensional and three-dimensional models can be useful to scientists, and that space scientists find models useful because the objects they study are usually so big and far away.

What's in the Sky?	Estimated Time
Brainstorming: What's in the sky?	15 minutes
Drawing sky objects	25 minutes
Introducing the concept of a model	15 minutes
Posting key concepts	5 minutes
TOTAL	**60 minutes**

What You Need

For the class
- ❑ *optional:* books with illustrations or posters of space
- ❑ sentence strips to record 6 key concepts
- ❑ wide-tip felt pen
- ❑ a chalkboard, white board, chart, or overhead projector
- ❑ pushpins or stapler to attach student drawings to a wall or bulletin board
- ❑ a model car, doll, doll furniture, model airplane, or other handy model to use to introduce the idea of models

Unit Goals

Size: Some sky objects are relatively small, and some are huge.

Distance: Some objects are relatively close to Earth and some are very far away.

Distance of sky objects from us affects their apparent size: Large objects appear small when far away.

TEACHER CONSIDERATIONS

TEACHING NOTES

Time Management: Although your students might enjoy making drawings for the whole session, stop them at least 20 minutes before class ends. You'll need this time to have a discussion about models and review key concepts introduced in this session.

Key Vocabulary

Science and Inquiry Vocabulary

Evidence

Scientific Explanation

Model

Scale Model

Prediction

Scientist

Three Dimensional (3-D)

Two Dimensional (2-D)

Space Science Vocabulary

Atmosphere

Satellite

Orbit

Diameter

Sphere

System

For each student:

- ❑ drawing paper (8½ x 11 inch paper is fine)
- ❑ colored pens, pencils, markers, crayons, or other drawing supplies

Getting Ready

1. **Student Sky Observations.** Prior to this lesson, you may want to ask your students to look at the sky at night and during the day to see what they can observe.

2. **Set out any books with illustrations or posters** you have of space for students to use as references.

3. **Have paper and drawing equipment handy.**

4. **Choose a wall or bulletin board to post your students' drawings** where they can remain during Unit 1.

5. **Write the following six key concepts on sentence strips** and have them ready to post during the session under the heading, *What We Have Learned About What Scientists Do.*

Scientists use models to help understand and explain how things work.
Space scientists use models to study things that are very big or far away.
Models help us make and test predictions.
Every model is inaccurate in some way.
Models can be 3–dimensional or 2–dimensional.
A model can be an explanation in your mind.

Unit Goals

Size: Some sky objects are relatively small, and some are huge.

Distance: Some objects are relatively close to Earth and some are very far away.

Distance of sky objects from us affects their apparent size: Large objects appear small when far away.

TEACHER CONSIDERATIONS

TEACHING NOTES

Time Management: The subject of what's in the sky and space usually elicits excitement, questions, and rich discussion. If possible, allow extra time for this initial discussion. Otherwise, let students know that they will have time during future sessions to discuss this exciting subject.

Science Class: Students sometimes bring up such questions as, "Where is heaven?" during discussions about space. Because this is a science class, try to keep students focused on a scientific perspective, and encourage them to bring up other questions in a more appropriate context.

What One Teacher Said

"The students are eager to share their thoughts and are not inhibited at all to share some things about space they have seen on TV! They are also asking questions that I am holding off on answering because I know they will be able to discover the answers further into the unit."

Key Vocabulary

Science and Inquiry Vocabulary

Evidence

Scientific Explanation

Model

Scale Model

Prediction

Scientist

Three–Dimensional (3-D)

Two–Dimensional (2-D)

Space Science Vocabulary

Atmosphere

Satellite

Orbit

Diameter

Sphere

System

Brainstorming What's in the Sky?

1. Introduce the brainstorm. Say that space scientists observe the sky. Tell them that you'll now make a list of things the students have observed in the sky. Write "Objects We Can See in the Sky" on the chalkboard or chart, and write two subtitles, "Night" and "Day," under it.

2. Brainstorm objects in the night sky. First, ask, "What have you seen in the sky at night using only your eyes, not telescopes or binoculars?" Record their responses under the "Night" heading. Accept any objects that students have seen. If someone mentions an object that can't actually be seen with the naked eye, such as Pluto, start a new column, "Other Sky Objects," and put Pluto there, explaining that it is very far away and can be seen only with telescopes. (The planets we can see with the unaided eye are Mercury, Venus, Mars, Jupiter, and Saturn.)

3. Brainstorm objects in the daytime sky. Next, ask students to brainstorm objects they have seen in the daytime sky. Record these in the column headed "Day."

4. Brainstorm objects in the sky that can't be seen with the naked eye. Now ask them to brainstorm objects in the sky or in space that they have heard about, but that can't be seen from Earth with the naked eye. Record these in the third column, headed "Other Sky Objects." Depending on your students, you may get a long list, including such exciting objects as Pluto, galaxies, supernovas, and black holes. Point out that space scientists gather evidence to try to understand and explain all these things.

Drawing Sky Objects

1. Draw sky objects. Congratulate your students on being good observers of the sky. Ask them to make drawings of sky objects. These can be from the class lists or can be other sky/space objects they think of. Each student should draw at least one object. Show students the available paper and drawing supplies, and point out any books or illustrations they can use as resources.

2. Draw only one object per page. Tell them that they may draw only one space object per page. This means that if they draw the Moon on their paper, they may not draw a star on the same paper. Have them begin.

Unit Goals

Size: Some sky objects are relatively small, and some are huge.

Distance: Some objects are relatively close to Earth and some are very far away.

Distance of sky objects from us affects their apparent size: Large objects appear small when far away.

TEACHER CONSIDERATIONS

TEACHING NOTES

Evidence of the Moon in the day: Many students are not aware that the Moon can be seen during the day at certain times of the month. If this question comes up, don't tell them that it can be seen. Instead, challenge them to use evidence to answer the question, Can anyone spot the Moon during the day?

UFOs and ETs: During the brainstorm, students may bring up UFOs or extraterrestrials. If so, tell them that some people have seen lights or other objects in the sky that haven't been identified. This is why they are called "unidentified flying objects," or UFOs. Some people believe UFOs are spaceships from another planet, but they have not presented the kind of testable evidence scientists need to verify their claims. For more information on this topic, see page 55 of the *Background* section.

You could present the optional role play about a "flying saucer" provided on page 109 in Session 1.1. However, although this is an exciting topic, refrain from discussing it at length right now.

Chart with some sample responses
Objects we can see in the sky:

Day	Night	Other Sky Objects
Sun	Moon	Pluto
clouds	stars	Uranus, Neptune
Moon	bats	supernovas
birds	clouds	galaxies
airplanes	airplanes	asteroids
rainbows	shooting stars	black holes

Satellite: An object that orbits a larger object. Some satellites, like the Moon, are natural. Artificial satellites are made by people. Note: Sometimes we can see satellites reflecting sunlight.

Meteor: The bright streak of light made by a bit of dust or rock heating up white hot as it enters the atmosphere at high speed. Meteors are also called shooting stars.

PROVIDING MORE EXPERIENCE

Go outside: Take your class outside to find as many sky objects as they can: Moon, birds, planes, Sun, etc. Remind students not to look directly at the Sun.

Sky Scavenger Hunt: To encourage sky-gazing, you may want to assign your students to do a sky "scavenger hunt" as homework. Briefly introduce the following sky objects and challenge them to spot as many as possible: Sun, star, Moon, meteor, planet, satellite. Also ask students to record if they saw each object high or low in the sky, and whether they spotted it during the day or night. Allow time in the next session to discuss what they've seen.

3. Collect drawings. After class, hang your students' work somewhere visible to the class. Have it remain there for the rest of the unit.

Introducing the Concept of a Model

1. Point out that drawings are not the same size as the real objects. Compliment them on their fine work, and then humorously point out that most of their drawings are not the same size as the real objects. For example:

- Call attention to a large drawing of a bird and a smaller drawing of the Sun, and say, "Look here—until now I didn't know that a bird is larger than the Sun."
- Say, "I didn't know that a rocket is the same size as a supernova."
- Say, "Look — a black hole is about the same size as my head."

2. Define models. Assure them that you know their drawings weren't supposed to be the sizes of the real sky objects—they are actually *models* of real objects. Say that a model is something that shows or helps explain what the real thing is like.

3. Define two- and three-dimensional models. Tell students their drawings are *two-dimensional* (2–D) models. This means they have the dimensions of height and width. Hold up a model car (or whatever model you chose) and say that some models are *three-dimensional* (3–D). The car has three dimensions: height, width, and depth.

4. Point out inaccuracy of models. Emphasize that good 2–D or 3–D models are like the real thing in some way(s), but no model is *exactly* the same as the real thing. Ask, "What are some ways the model car is *not* exactly the same as a real car?" [It's smaller, has no motor, the doors don't open, the tires are metal, it doesn't have gas in it, it has no lights, it can't move under its own power, and so on.]

5. Introduce idea of mental models. Mention that we can have a picture in our minds about how something works. This is a kind of model too, someone's explanation in their mind of how something looks or works.

6. Explain that scientists often use models. Say scientists use all kinds of models to explain things they have observed, show how they think things in the natural world work, make predictions, or learn more about things they can't look at directly. Space scientists often use models, because many of the things they study are large and far away.

Unit Goals

Size: Some sky objects are relatively small, and some are huge.

Distance: Some objects are relatively close to Earth and some are very far away.

Distance of sky objects from us affects their apparent size: Large objects appear small when far away.

TEACHER CONSIDERATIONS

TEACHING NOTES

One object per page: The reason for having students draw only one space object per page is that their drawings will be referred to as "models" later in the session. Also, an optional activity in a later session on scale is to have students sort their drawings by the real sizes of the objects, and it's difficult to sort papers with multiple objects.

2-D and 3-D: In the discussion of what makes something 2-D or 3-D, additional props may be useful. For example, you could use a 3-D cube and a 2-D square that are the same size, or a 3-D sphere and a 2-D disk to help illustrate the point.

Key Vocabulary

Science and Inquiry Vocabulary

Evidence

Scientific Explanation

Model

Scale Model

Prediction

Scientist

Three–Dimensional (3-D)

Two–Dimensional (2-D)

Space Science Vocabulary

Atmosphere

Satellite

Orbit

Diameter

Sphere

System

Adding to the Concept Wall

1. Add to concept wall. Read the key concepts about models aloud, one at a time, and post them on the concept wall. Tell students that they do not need to remember all of these now, but they will be using models and referring to these key concepts during the coming activities.

Scientists use models to help understand and explain how things work.
Space scientists use models to study things that are very big or far away.
Models help us make and test predictions.
All models are inaccurate in some way.
Models can be 3–dimensional or 2–dimensional.
A model can be an explanation in your mind.

Unit Goals

Size: Some sky objects are relatively small, and some are huge.

Distance: Some objects are relatively close to Earth and some are very far away.

Distance of sky objects from us affects their apparent size: Large objects appear small when far away.

TEACHER CONSIDERATIONS

OPTIONAL PROMPTS FOR WRITING OR DISCUSSION

You may want to choose one or more of the prompts below for science journal writing during class or as homework. Or they could be used for a discussion or during a final student sharing circle in which each student gets a turn to share.

- Name a model (such as a doll or model plane) that you have at home or have seen before.
- Describe in what ways a certain model is like the real thing (accurate) and in which ways it is not like the real thing (inaccurate).
- List the accuracies and inaccuracies of a model on a T-chart.

Key Vocabulary

Science and Inquiry Vocabulary

Evidence

Scientific Explanation

Model

Scale Model

Prediction

Scientist

Three–Dimensional (3-D)

Two–Dimensional (2-D)

Space Science Vocabulary

Atmosphere

Satellite

Orbit

Diameter

Sphere

System

SESSION 1.3
Measuring Sizes of Objects

Overview

The session begins with an introduction (or review) of how to measure the lengths of objects using metric units. During your demonstration, students measure a string representing the wingspan of a sky object, one of the biggest birds (an albatross). In teams of two, students then measure the wingspan of the smallest bird (a hummingbird) and of four satellites.

This activity reinforces the key concepts about models introduced in Session 1.2 and gives students an opportunity to practice measurement with metric units. Measuring and comparing the actual sizes of birds and satellites sets the stage for later sessions in which students measure the sizes and distances of much larger space objects using scale models.

Measuring Sizes of Objects	Estimated Time
Practice measuring with metric units	20 minutes
Measuring a bird and four satellites	30 minutes
Discussing the measurement activity	10 minutes
TOTAL	**60 minutes**

What You Need

For the class

- ❑ overhead projector or computer with large screen monitor/LCD projector
- ❑ overhead transparency from the transparency packet or CD-ROM file of *Measuring Length in Metric Units*
- ❑ 1 measuring tape with metric units
- ❑ 1 piece of string* 3 meters and 63 centimeters long (length of the wingspan of an albatross)
- ❑ 5 indelible markers in different colors to color-code string lengths
- ❑ 1 piece chart paper

**Note on String: String should be as sturdy and non stretchy as possible. Do not use yarn, because it stretches more than string, and will result in more inaccurate measurements.*

Unit Goals

Size: Some sky objects are relatively small, and some are huge.

Distance: Some objects are relatively close to Earth and some are very far away.

Distance of sky objects from us affects their apparent size: Large objects appear small when far away.

TEACHER CONSIDERATIONS

CD-ROM NOTES

Wait to use "What's Up?" interactive: On the CD-ROM there is an interactive called "What's Up?" that students can use to independently explore sizes and other aspects of sky objects. However, *wait until after Session 1.6 to make this interactive available to students.* This way they will discover the relative sizes for themselves first, and the CD-ROM will serve as reinforcement, as well as enrichment.

What One Teacher Said

"My students loved this! It was a great exercise in measuring accurately.... The students are more confident with metrics even though that wasn't the main intention of the lesson. Using metrics for a real (scientific) purpose seems to have an impact. By the time they finished this lesson, they seemed to have a real grasp of the size differences between these sky objects."

Key Vocabulary

Science and Inquiry Vocabulary

Evidence

Scientific Explanation

Model

Scale Model

Prediction

Scientist

Three–Dimensional (3-D)

Two–Dimensional (2-D)

Space Science Vocabulary

Atmosphere

Satellite

Orbit

Diameter

Sphere

System

For each group of four students

- ❑ 2 student data sheets: *Measuring Hummingbird and Satellites,* from the student sheet packet
- ❑ 5 pieces of string* in different lengths (see *Getting Ready* which follows)
- ❑ 5 plastic fold-top sandwich bags
- ❑ 1 paper clip
- ❑ 2 measuring tapes with metric units

Getting Ready

1. **Arrange for the appropriate projector format** (computer with large screen monitor, LCD projector, or overhead projector) to display images to the class.

2. **If you will not be using the CD-ROM, make an overhead transparency of *Measuring Length in Metric Units.***

3. **Make a copy for each pair of students of the *Measuring Hummingbird and Satellites* student sheet.**

4. **Cut one piece of string 3 meters, 63 centimeters long** to represent the actual wingspan of an albatross for a class demonstration. Label it with a piece of masking tape or put it in a labeled sandwich bag and set it aside for your introduction.

5. **Make a chart** like the one below on the board, or on an overhead transparency, or on chart paper:

Class Chart of Bird and Satellite Measurements

Object	Measurement
Hummingbird Wingspan	
Satellite #1: (*Sputnik*)	
Satellite #2: (Ocean Study)	
Satellite #3: (Space Experiments)	
Satellite #4: (Swift)	

Unit Goals

Size: Some sky objects are relatively small, and some are huge.

Distance: Some objects are relatively close to Earth and some are very far away.

Distance of sky objects from us affects their apparent size: Large objects appear small when far away.

Key Vocabulary

Science and Inquiry Vocabulary

Evidence

Scientific Explanation

Model

Scale Model

Prediction

Scientist

Three–Dimensional (3-D)

Two–Dimensional (2-D)

Space Science Vocabulary

Atmosphere

Satellite

Orbit

Diameter

Sphere

System

6. Prepare the lengths of string. Each team of four students will need five pieces of string in different lengths. These will represent the lengths of five sky objects. If possible, get a volunteer to help you cut, label, and organize the string pieces.

a. Cut string. Decide on how many teams you will have, and cut enough of each length of string so that each team will have one each of the following:

Hummingbird: 10 cm
Satellite #1: 58 cm
Satellite #2: 150 cm
Satellite #3: 47 cm
Satellite #4: 127 cm

b. Mark string pieces with colors so teams can tell them apart. Using indelible markers, mark one end of each piece of string with a different color. Make a key to post on the wall for students to refer to during the activity. For example:

Hummingbird: red
Satellite #1: blue
Satellite #2: green
Satellite #3: purple
Satellite #4: orange

c. Put one string of each color in a sandwich bag for each team. Use an indelible marker to label the bags "bird," #1, #2, #3, and #4.

d. Do the same for each team.

Practice Measuring with Metric Units

1. Introduce metric units of measure. Tell the class they will be measuring the sizes of some sky objects today to see how big they are. Say that they may be used to measuring things in inches, feet, yards, and miles. The United States and a few other countries use these units. However, most of the world—and all scientists in the world—use *metric units*. Tell students that they will be using metric units.

2. Show that 1 meter equals 100 centimeters. Hold up a metric measuring tape or meter stick. Show how long a meter is. Say that *cent* means "100." Ask them to imagine the meter divided into 100 pieces. Tell them that's what centimeters are. Have them hold their fingers about 1 centimeter (1 cm) apart.

Unit Goals

Size: Some sky objects are relatively small, and some are huge.

Distance: Some objects are relatively close to Earth and some are very far away.

Distance of sky objects from us affects their apparent size: Large objects appear small when far away.

TEACHER CONSIDERATIONS

TEACHING NOTES

Adjusting for Student Experience: If your students have already had experience with measurement using metric units, you may choose to skip or abbreviate the practice steps, and move directly to measuring the albatross wingspan, Step #7 on page 128.

If this is your students' first experience with metric measurement, this session could serve as an introduction to metrics, but it will take longer. You might consider breaking it into two sessions. The first could focus on how to measure with metrics, and the second on the measuring of the bird wingspan and satellites.

U.S. units of measure:
12 inches = 1 foot
3 feet = one yard
1,760 yards = 1 mile

Metric units of measure:
10 millimeters (mm) = 1 centimeter (cm)
100 centimeters (cm) = 1 meter (m)
1,000 meters (m)= 1 kilometer (km)

Key Vocabulary

Science and Inquiry Vocabulary

Evidence
Scientific Explanation
Model
Scale Model
Prediction
Scientist
Three–Dimensional (3-D)
Two–Dimensional (2-D)

Space Science Vocabulary

Atmosphere
Satellite
Orbit
Diameter
Sphere
System

3. Show that 1 meter equals 1,000 millimeters. Ask them to imagine dividing a centimeter into 10 pieces. There are 1,000 millimeters in a meter. Point out the *millimeters (mm)*. Say that *mille* means "1000." Ask them to imagine the meter divided into 1,000 pieces. Tell them that's what millimeters are.

4. Use *Measuring Length in Metric Units* to show how to use a measuring tape. Show *Measuring Length in Metric Units* to the class. Say that some rulers and measuring tapes have both kinds of units on them. Point out the centimeters and inches. Caution students to be sure to use centimeters, which are the smaller units.

5. Point out common measuring mistakes. Using the overhead transparency, point out some common measuring mistakes:

- There is no "0" on the ends of rulers and measuring tapes. Some students measure beginning where the number "1" is instead of the end, where the "0" would be.

- Some students get confused by metal tabs at the end of a measuring tape. They measure starting from where the metal tab begins, instead of the end of the tape.

6. Measure something shorter than a meter. Ask a student to help demonstrate how to measure a book. Emphasize:

a. Put one end of the tape at one edge of the book. This should be the end of the tape with the number "1." The very end of the tape should be even with the edge of the book.

b. Note the number on the tape nearest to the other end of the book.

c. Record this number. Tell the student to write the length of the book on the board. Write both the word "centimeters" and the abbreviation "cm" after the number.

7. Introduce challenge of measuring the wingspan of an albatross. Tell them that one of the things they're going to be measuring is the wingspan of one of the largest birds there is, the albatross.

8. Define *prediction*. Explain that scientists often make thoughtful guesses beforehand about what they will find out. Say that a prediction is a guess about what they will find out, and the guess is usually based on earlier evidence.

Unit Goals

Size: Some sky objects are relatively small, and some are huge.

Distance: Some objects are relatively close to Earth and some are very far away.

Distance of sky objects from us affects their apparent size: Large objects appear small when far away.

TEACHER CONSIDERATIONS

MEASURING LENGTH IN METRIC UNITS

Session 1.3 Transparency

Measuring Length in Metric Units

1. Make sure you are using centimeters and meters, not inches and feet.

Metrics (use this!)

Inches and Feet (don't use this!)

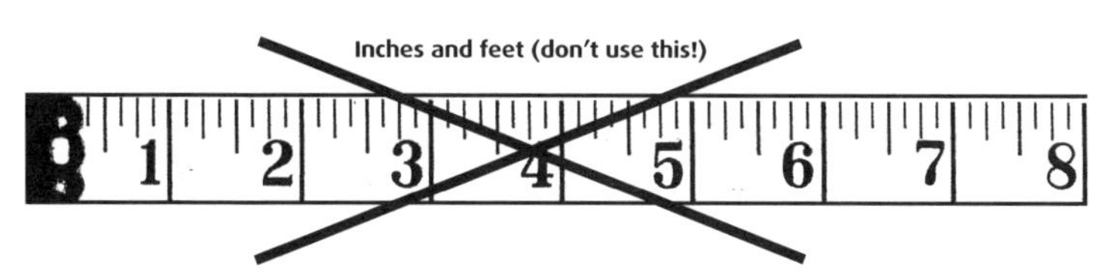

2. Put the end with the smaller numbers at one end of what you are measuring.

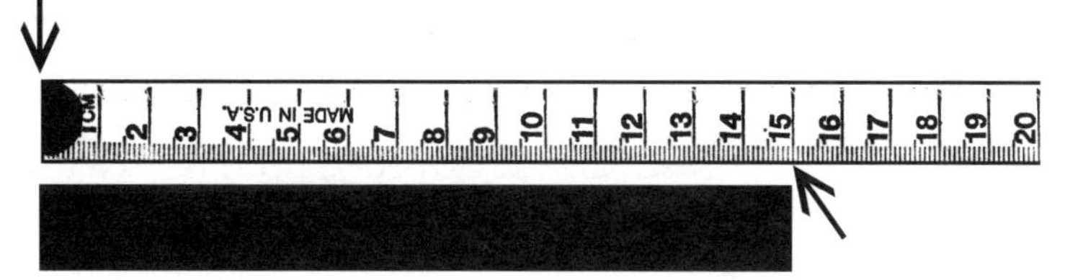

3. Look at what number is at the other end of what you are measuring. That's the measurement.

4. 100 centimeters = 1 meter. If you measure 110 centimeters, you can also write it as 1 meter and 10 centimeters.

9. Predict length of an albatross. Have your students spread their arms apart to help predict how long they think it will be from the tip of one wing of an albatross to the other.

10. Measure the albatross. Tell them you don't have the actual bird for them to measure, but you have a string that is the same length. Ask three students to help measure the model albatross wingspan as the class watches.

a. Have two students hold either end of the string representing the length of an albatross's wingspan and stretch it out horizontally.

b. Have the third student put one end of the measuring tape at one end of the string.

c. Demonstrate how to mark the 1 meter point with a finger on the string. Then, show how to pick up the measuring tape, and measure the next meter from that point. Continue doing this until you reach the end of the string, about 3meters and 63 centimeters.

11. Record metric measurements. On the board, show one or more of the following ways they can record their measurement:
3 meters, 63 centimeters
3 m 63 cm
363 centimeters
363 cm

Students Measure a Bird and Four Satellites

1. Introduce *satellites*. Ask students what they know about *satellites.* [Satellites made by people travel around the Earth in space. They take photos and other images of Earth, relay cell phone/TV/pager messages.] Explain that because they are made by people, they are sometimes called artificial satellites. The Earth's Moon, on the other hand, is a natural satellite.

2. Define *orbit*. Tell students that people have made many satellites and sent them up in the sky to *orbit* the Earth. Ask them to help define what *orbit* means. [Orbit means to move around another object in space.] An orbit can also be the name for the path taken by one object circling around another object.

Unit Goals

Size: Some sky objects are relatively small, and some are huge.

Distance: Some objects are relatively close to Earth and some are very far away.

Distance of sky objects from us affects their apparent size: Large objects appear small when far away.

MEASURING A HUMMINGBIRD AND FOUR SATELLITES

TEACHING NOTES

Satellites and Satellite Dishes: Make sure that students understand the distinction between satellites and satellite dishes— satellite dishes are **on Earth**; they send and receive messages from satellites in orbit.

A satellite is actually any object that orbits another. The Earth is a satellite of the Sun, and the Moon is a satellite of the Earth.

Modeling the wrong way: If students need a review of the measuring procedure, tell them you're going to go over the procedure by doing everything the wrong way. Model every measuring mistake or uncooperative behavior you can think of (for example, you could act bossy, take the measuring tape away from your partner, measure in inches rather than centimeters, or begin your measurement from the 1 cm mark, rather than from the end of the tape). Have students point out what you did wrong, and tell you what would be the "right way."

This strategy can be a very effective step-by step review of the procedure and what not to do. The portrayal of common student mistakes is also very entertaining for students. They love to watch the teacher doing things the wrong way, especially if you exaggerate your errors for comedic effect. Some teachers prefer to tell their students that they do not need to raise their hands during this exercise, but can simply call out.

Name:____________________

Measuring a Hummingbird and Four Satellites

Five Sky Objects	How long is it in centimeters? (cm)
Hummingbird Wingspan	
Satellite #1: *Sputnik*	
Satellite #2: Ocean Study	
Satellite #3: Space Experiments	
Satellite #4: *Swift*	

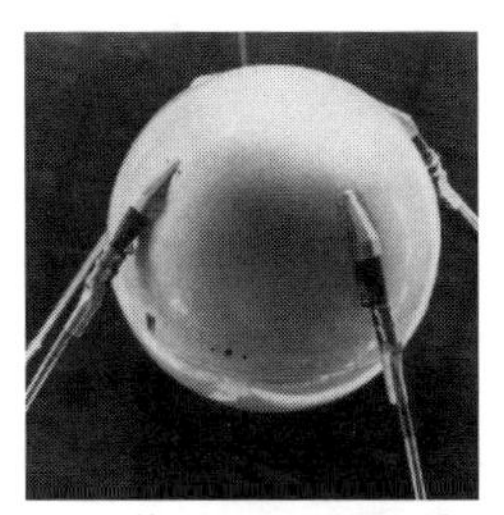

Satellite #1: *Sputnik*

Satellite #2: Ocean Study

Satellite #3: Space Experiments

Satellite #4: *Swift*

3. Introduce the measurement activity. Hand out a *Measuring a Hummingbird and Four Satellites* sheet to each pair of students and give them a chance to look at the photos of satellites. Tell your students that satellites come in different sizes. They're going to measure the length of four satellites and one small bird, the hummingbird, for comparison. They will not be measuring the actual bird or satellites, but pieces of string that will serve as models of the lengths of the real objects.

4. Predict lengths of smallest and largest satellites. Ask students to predict how long they think the largest and smallest satellites might be.

5. Share materials; measure and record results. Tell them that they'll work in teams of two, sharing a student sheet and measuring tape, and that they'll share the pieces of string with all the people at their table. Show the string pieces in labeled bags, and explain how they are color-coded. Point out the color key on the wall.

6. Begin measuring. Show them where they will record their measurements on the *Measuring a Hummingbird and Four Satellites* sheets. They can start with any of the five, but the hummingbird and satellites #1 and #3 are easier to start with. Have them begin.

Discussing the Measurement Activity

1. Organize and collect materials. When teams have finished, tell them to put their table's string pieces back into the appropriate bags. Collect all materials, but have students keep their student sheets.

2. Debrief the measurement data. Regain the attention of the class and start by asking a few students for their hummingbird measurements. If most of the class agree on an approximate measurement, such as 10 cm, record it on the class chart you made earlier.

3. Discuss varying measurement data. The students' results for the satellite measurements are likely to vary. If there is variation, ask, "What might have caused you to get different measurements? [The string stretches, so it's hard to measure exactly. Maybe whoever cut the string didn't cut them all the same lengths. Or maybe some students didn't start measuring with the end of the measuring tape, or made other errors.]

Unit Goals

Size: Some sky objects are relatively small, and some are huge.

Distance: Some objects are relatively close to Earth and some are very far away.

Distance of sky objects from us affects their apparent size: Large objects appear small when far away.

TEACHER CONSIDERATIONS

TEACHING NOTES

Measurements: It is okay if the averages of your students' measurements do not exactly match the actual sizes of the satellites on the sample chart.

PROVIDING MORE EXPERIENCE

Some teachers extend the activity with more mathematics:

- Discuss "outliers" and vote on which numbers to toss out before averaging the class data (outliers are results that are widely divergent, on either the large or small side).
- Have students average the class measurements with calculators. Then have them figure out the median and the mode.
- Make a graph representing the measurements for each sky object.

ASSESSMENT OPPORTUNITY

Critical Juncture/Embedded Assessment: As student pairs report their measurement data, you may notice that some measurements are very different from the others. If so, ask all pairs to hand in their recording sheets. When you've reviewed their work after class, check with them to find out where they went wrong and provide more practice. It's important for all students to have confidence in their measurement abilities, because more measurement activities follow in the next sessions.

PROVIDING MORE EXPERIENCE

More practice: If students need it, provide more practice measuring objects around the room, such as desks, doors, pens, and so on.

Story with metric practice: To deepen their familiarity with metric units, tell a story aloud with a variety of metric measurements in it. Tell students that each time you read a measurement, they should show with their hands approximately how long that is. For example, "There once was a girl who was 1 meter tall. Her hair was 10 centimeters long. One day. . ." Have them practice estimating and visualizing metric system measurements, and then use a metric measuring tape to check their estimates. Give them a signal to use if a measurement you mention in the story is too long to show with their arms.

What Some Teachers Said

"The students were amazed at how big the wingspan of an albatross really was. They also discovered that not all satellites are as big as they had originally thought."

"Using models (the strings) is just as effective as having the real thing, and the students were able to understand this."

Key Vocabulary

Science and Inquiry Vocabulary

Evidence

Scientific Explanation

Model

Scale Model

Prediction

Scientist

Three–Dimensional (3-D)

Two–Dimensional (2-D)

Space Science Vocabulary

Atmosphere

Satellite

Orbit

Diameter

Sphere

System

4. Record approximate average of group data on the chart. As long as the measurements are roughly similar, it's fine for this activity. Tell them that you won't count measurements that are vastly different from the others. Make an approximate average of the group data, and record it on the chart. As you record, review metric units by asking questions such as, "Is 47 cm more or less than a meter?" or "How many meters are in 150 cm?"

Your finished chart should look something like this:

Object	Measurement
Hummingbird wingspan	10 cm
Satellite #1 *Sputnik*	58 cm
Satellite #2: Ocean Study	150 cm
Satellite #3: Space Experiments	47 cm
Satellite #4: Swift	127 cm

5. Say that they have gathered *evidence* about some sky objects. Point out the concepts about evidence on the concept wall, and say that they have used models to gather evidence about two birds and four satellites and that they now know the real sizes of these sky objects. Ask them to compare the sizes of hummingbirds, albatrosses, or satellites with the sizes of the drawings of any birds or satellites from Session 1.

6. They will gather evidence of the real sizes of the Earth, Moon, and Sun. Say that in the next session they will use models to gather some evidence about how big the Earth, Moon, and Sun really are.

Unit Goals

Size: Some sky objects are relatively small, and some are huge.

Distance: Some objects are relatively close to Earth and some are very far away.

Distance of sky objects from us affects their apparent size: Large objects appear small when far away.

TEACHER CONSIDERATIONS

OPTIONAL PROMPTS FOR WRITING OR DISCUSSION

You might use the first prompt below for a sharing circle in which each student gets a turn to share. The second prompt could be used for science journal writing during class or as homework.

- Did any of the measurements surprise you? Why?

- Use a metric ruler to measure ten objects in your classroom or bedroom. Include one very small object, and one large one.

OBJECT	MEASUREMENT

Key Vocabulary

Science and Inquiry Vocabulary

Evidence

Scientific Explanation

Model

Scale Model

Prediction

Scientist

Three–Dimensional (3-D)

Two–Dimensional (2-D)

Space Science Vocabulary

Atmosphere

Satellite

Orbit

Diameter

Sphere

System

SESSION 1.4

How Big Are the Earth, Moon, and Sun?

Overview

Session 1.4 begins with a review of models. Students are introduced to scale models, and learn that space scientists find scale models useful because the objects they study are usually so big and far away.

Next, students predict the relative sizes of the Earth, Moon, and Sun. After a review of measurement and an introduction to kilometers, students measure the diameters of models of the Earth, Moon, and Sun using special scale rulers on which 1 mm represents 3,000 km. They also measure how many "Earths" fit across the Sun's diameter. In a class discussion of their measurements, the class concludes that the Moon is very big, the Earth is huge, and the Sun is super huge. A look at 3-D scale models gives them a sense of the stunning differences in size between these objects. They also learn that the Sun is a star, and is average-sized for a star.

In Session 1.5, there will be an opportunity to reflect on the data they have gathered, and to discuss such questions as, "Why does the Sun look so small if it's really so big?"

How Big Are the Earth, Moon, and Sun?	Estimated Time
Reviewing models and introducing scale models	10 minutes
Measuring scale models to learn the real sizes of the Earth, Moon, and Sun	30 minutes
Discussing the measurements	10 minutes
Comparing the sizes	10 minutes
TOTAL	**60 minutes**

What You Need

For the class

- ❑ a paper cutter or scissors to cut rulers off the edges of the *Measuring the Diameters of the Earth, Moon, and Sun* student sheets
- ❑ 1 tiny piece of white paper (5mm x 5mm) or a cake decoration about 1 mm in diameter
- ❑ a piece of blue paper 1cm x 1cm or a cake decoration about 4 mm in diameter
- ❑ a bit of clear tape
- ❑ enough newspaper to wad into a "Sun" ball 44 cm in diameter (or, better yet, a 44 cm diameter ball or balloon)
- ❑ 1 roll masking tape to tape the 44 cm "Sun" ball together
- ❑ 6 sentence strips for the concept wall

Unit Goals

Size: Some sky objects are relatively small, and some are huge.

Distance: Some objects are relatively close to Earth and some are very far away.

Distance of sky objects from us affects their apparent size: Large objects appear small when far away.

TEACHER CONSIDERATIONS

TEACHING NOTES

The scale model used in Session 1.4: The scale model used to compare the sizes of the Earth, Moon, and Sun includes very small models of the Moon and Earth–1 mm and 4 mm respectively. Although it is a bit inconvenient to measure such small models, the advantage is that the same scale can be used in Session 1.8 to study the distances between the Earth, Moon, and Sun.

Key Vocabulary

Science and Inquiry Vocabulary

Evidence
Scientific Explanation
Model
Scale Model
Prediction
Scientist
Three–Dimensional (3-D)
Two–Dimensional (2-D)

Space Science Vocabulary

Atmosphere
Satellite
Orbit
Diameter
Sphere
System

- ❑ wide-tip felt pen
- ❑ a 22–cm cardboard strip
- ❑ a toy car or other model(s) you used in Sessions 1.1 and 1.2

For each pair of students

- ❑ 1 circle of paper 44 cm in diameter (yellow or white paper if possible); see "Getting Ready," below

For each student

- ❑ 1 copy of *Measuring the Diameters of the Earth, Moon, and Sun* student sheet, from the student sheet packet

Getting Ready

1. Photocopy a *Measuring the Diameters of the Earth, Moon, and Sun* sheet for each student.

2. Cut out and trim scale rulers. Check to see that the scale on the rulers on the edges of the student sheets have not been distorted by the photocopier. If they have been, you may need to cut out the ruler from the copy master, place it in the middle of a page, and recopy it. To make the rulers easier for students to use, trim the two ends so the marks for 0 km and 600,000 km are as close as possible to the edge of the ruler. Also, trim the long edge of the ruler with the scale markings to remove the blank border. *Optional: Laminate the rulers.*

3. Make two-dimensional paper Sun models. If possible, have a volunteer help you with this task, as it can be time consuming. On a large sheet of paper (preferably yellow or white), measure and cut out a 44–cm–diameter circle for each pair of students. To do this, get a strip of cardboard or file folder material at least 24 cm long. With a pushpin, make two holes 22 cm apart. Stick the pushpin through one of the holes and into the center of the yellow paper (you can put a little piece of corrugated cardboard below for the pushpin to stick into). Stick a pencil or pen through the other hole, and draw the circle, using a 22–cm cardboard strip as the radius. Rather than using the cardboard strip method each time, you might want to trace around the first disk to draw the others.

4. Draw the diameter on each paper Sun. Use a meter stick to draw a line across the circle. Choose a place on your cut-out circle where the diameter is exactly 44 cm. Students will measure the diameter along this line, so it's important that it be the correct length. It's okay if the rest of the model Sun isn't a perfect circle.

Unit Goals

Size: Some sky objects are relatively small, and some are huge.

Distance: Some objects are relatively close to Earth and some are very far away.

Distance of sky objects from us affects their apparent size: Large objects appear small when far away.

TEACHER CONSIDERATIONS

Key Vocabulary

Science and Inquiry Vocabulary

Evidence

Scientific Explanation

Model

Scale Model

Prediction

Scientist

Three–Dimensional (3-D)

Two–Dimensional (2-D)

Space Science Vocabulary

Atmosphere

Satellite

Orbit

Diameter

Sphere

System

5. Prepare three-dimensional models of the Earth, Moon, and Sun. The models described here are made from paper. However, if you have cake decorations the right size for the Moon and Earth models, or a large balloon or ball about 44 cm for the Sun model, you can use those instead. Keep the models out of sight until it's time to reveal them to the class.

Moon = 1–mm diameter. Tear a tiny piece of white paper about 5 mm X 5 mm. Roll it into a tiny round shape about 1 mm in diameter. Use clear tape to attach it to a colored piece of paper. Label it "Moon."

Earth = 4–mm diameter. Tear a piece of blue paper about 1 cm X 1 cm. Roll it into a tiny round shape about 4 mm in diameter. Tape it to a white piece of paper and label it "Earth."

Sun = 44–cm diameter. Wad up newspaper pages into a ball roughly 44 cm in diameter. Use masking tape to tighten it up and make it round. Label it "Sun" with large lettering. Keep this model out of sight until it is time to reveal it at the end of the class session.

6. Write the six key concepts on sentence strips, and have them ready to post at the end of the session:

Some objects in the sky, such as the Sun, Moon, stars, and planets, are very large.
Other objects in the sky, such as birds, satellites, and airplanes, are relatively small.
The Earth is very large.
The Moon is very large, but not as large as the Earth.
The Sun is super huge compared with the Earth.
The Sun is a star. Compared with other stars, it is medium–sized.

Unit Goals

Size: Some sky objects are relatively small, and some are huge.

Distance: Some objects are relatively close to Earth and some are very far away.

Distance of sky objects from us affects their apparent size: Large objects appear small when far away.

TEACHER CONSIDERATIONS

TEACHING NOTES

Tips for Leading Good Discussions: Engaging students in thoughtful discussions is a powerful way to enrich their learning. Even the most experienced teacher can use a few reminders about how to lead a good discussion. Listed below are some general strategies to keep in mind while leading a class discussion:

- Ask broad questions (questions which have many possible responses) to encourage participation.
- Use focused questions sparingly (questions which have only one correct response) to recall specific information.
- Use wait time (pause about 3 seconds after asking a question before calling on a student).
- Give non-judgmental responses, even to seemingly outlandish ideas.
- Listen to student responses respectfully, and ask what their evidence is for their explanations.
- Ask other students for alternative opinions or ideas.
- Try to create a safe, non-intimidating environment for discussion.
- Try to call on as many females as males.
- Try to include the whole group in the discussion.
- Offer "safe" questions to shy students.
- Employ hand raising or hand signals to insure whole group involvement.
- Take time to probe what students are thinking.
- Consider your role as a collaborator with the students, trying to figure things out together.
- Encourage students to figure things out for themselves, rather than telling them the answer.

Key Vocabulary

Science and Inquiry Vocabulary

Evidence

Scientific Explanation

Model

Scale Model

Prediction

Scientist

Three–Dimensional (3-D)

Two–Dimensional (2-D)

Space Science Vocabulary

Atmosphere

Satellite

Orbit

Diameter

Sphere

System

Reviewing Models and Introducing Scale Models

1. Review models. Ask, "What is a model?" Have students refer to the concept wall as they review. [A model is an object that is used to help understand, predict, or explain how things work. A drawing or even an explanation in your mind can be a model.]

2. Define scale models. Hold up the model you used in Session 1.2. Say that although it's much smaller than a real car, this model looks like a car because someone measured every part and made each part smaller by the same amount. It is a *scale model* of a real car.

3. Compare scale models. If you had a model truck that was "shrunk down" exactly the same amount as the car, you could compare the sizes of the two models. You could use the models to see how the sizes of the real car and truck compare with each other.

4. Scale models of Earth, Moon, and Sun. Tell them that because the real sizes of the Earth, Moon, and Sun can't fit in the classroom, the class is going to measure some scale models to get a better understanding of how big these objects are, compared with one another.

5. Predict the relative sizes of Earth, Moon and Sun. Ask the class for their ideas about the sizes of the real Earth, Moon, and Sun: Are they all the same size? Different sizes? If they are different sizes, which one is biggest or smallest? How different are their sizes? Accept all answers.

6. Use 2-D models. Tell them the real Earth, Moon, and Sun are all shaped like balls, but to start with the class will use 2-dimensional disks or circles as models so their sizes can be measured more easily.

7. Hold up one of the 2-D paper Sun models. Ask, "If the Sun were this size, how big do you think the Moon would be? Bigger or smaller or the same? About what size would you guess?

Measuring Scale Models to Learn the Real Sizes of the Sun, Earth and Moon

1. Review millimeters, centimeters, and meters. Ask students to show with their fingers and/or arms their best approximations of the following metric units:

- 1 millimeter
- 1 centimeter
- 1 meter

Ask, "How many centimeters are in one meter?" [100.] "How many millimeters are in one centimeter?" [10.]

Unit Goals

Size: Some sky objects are relatively small, and some are huge.

Distance: Some objects are relatively close to Earth and some are very far away.

Distance of sky objects from us affects their apparent size: Large objects appear small when far away.

TEACHER CONSIDERATIONS

Key Vocabulary

Science and Inquiry Vocabulary

Evidence

Scientific Explanation

Model

Scale Model

Prediction

Scientist

Three–Dimensional (3-D)

Two–Dimensional (2-D)

Space Science Vocabulary

Atmosphere

Satellite

Orbit

Diameter

Sphere

System

2. Review kilometers. Remind students that the sheep, duck, and chicken they read about in Session 1 went 2000 meters up in the sky. Ask, "How many kilometers is that?" [2 km.] Remind them of the local landmark about two kilometers away.

3. Introduce the scale ruler. Pass out a paper scale ruler to each student. Caution students to treat them carefully, as they tear easily. Explain that these rulers look similar to the measuring tapes they used in Session 3, only the labels have been changed to help them figure out the real sizes of the Sun, Earth, and Moon by measuring scale models. Go over the markings on the scale ruler. Help students figure out what 2 mm would represent, and so on, with several examples.

1 mm = 3,000 km
1 cm = 30,000 km (this is 10 times longer than 3,000 km)
1 m = 3,000,000 km (this is too far for us to be able to imagine)

4. Demonstrate how to measure a diameter. Show the line you drew across the widest part of a flat paper Sun model and ask, "What is this part of a circle usually called?" [*Diameter.*] Demonstrate how to use the scale ruler to measure the diameter of the Sun straight from one side to another across the center of the disk.

5. Pass out a *Measuring the Diameters of the Earth, Moon, and Sun* sheet to each student. Ask them to find the Earth and Moon scale models drawn on the sheet. Emphasize that these scale models have all been "shrunk" down by the same amount as the paper Sun model. Ask students how the model Earth and Moon on the page compare to the paper model Sun. *The Sun is super huge compared with the Earth and Moon!*

6. Go over the questions on the sheet. Make sure they understand that they will use their rulers to measure the paper Sun model, and the 2-D models of the Earth and Moon pictured on their sheets, all of which are "shrunk down" by the same amount. They will record on their papers how many thousand of kilometers each object's diameter really is.

Unit Goals

Size: Some sky objects are relatively small, and some are huge.

Distance: Some objects are relatively close to Earth and some are very far away.

Distance of sky objects from us affects their apparent size: Large objects appear small when far away.

TEACHER CONSIDERATIONS

TEACHING NOTES

Scale Rulers: The scale rulers can be challenging for some students to use. For younger and/or less experienced students, you may want to provide extra practice measuring various lengths with the scale ruler.

MEASURING THE DIAMETERS . . .

0 km, 30,000, 60,000, 90,000, 120,000, 150,000, 180,000, 210,000, 240,000, 270,000, 300,000, 330,000, 360,000, 390,000, 420,000, 450,000, 480,000, 510,000, 540,000, 570,000, 600,000

Scale Ruler: 1 mm = 3,000 km; 1cm = 30,000 km

Measuring the Diameters of the Earth, Moon, and Sun

Practice with Scale Rulers

1. Use the scale ruler to measure how many kilometers this line represents:

   Write the number here: ______ km

2. Using the ruler, draw a line below representing fifteen thousand kilometers (15,000 km).

3. Using the ruler, draw a line below representing sixty thousand kilometers (60,000 km).

To measure the diameter of a circle, measure from one side to the other, going through the center of the circle.

4. Measure the diameter of the model of the Earth. ●
   Earth's diameter: ________

5. Measure the diameter of the model of the Moon. •
   Moon's diameter: ________

6. Measure the diameter of the model of the Sun (not on this page)
   Sun's diameter: ________

7. How many Earths could fit across the diameter of the Sun? __________

8. **Size Challenge**: Put the following objects in order from smallest to biggest:

Smallest: ________________ mountain

________________ Sun

________________ largest bird

________________ *Sputnik* satellite

________________ Moon

________________ Earth

Biggest: ________________ smallest bird

Student Sheet—Space Science Sequence 1.4

7. Brainstorm strategies for answering #7. They will work with a partner on question #7 to figure out how many Earths could fit across the diameter of the Sun. Ask them to discuss with their partner how to do this, then try one or more ways.

8. Explain logistics of sharing materials and begin. Tell students that the Sun models need to be shared by pairs of students, so they need to cooperate. Divide the class into table groups of four to six, pass out one paper Sun model to each student pair, and have them begin.

Discussing the Measurements

1. Discuss their measurements. When students have completed their measurements, regain their attention and ask for general comments on what they learned.

2. Report diameter of the Earth. Ask for their measurements of the diameter of the Earth. Their answers may vary, but should be in the ballpark of 12,000 km. [The real diameter is 12,750 km, but this can't be determined precisely with the model they have.] Acknowledge that, with the paper rulers and small scale model, it can be difficult to make exact measurements.

3. Earth is very large! Conclude that 12,000 km is 6,000 times the distance from your school to [name the local landmark that's two km away]. And that's only the diameter—not the distance *around* the Earth. So, the Earth is really huge!

4. Report diameter of the Moon. Ask for their measurements of the diameter of the Moon. Their answers may vary, but should be roughly 3,000 km. [The real diameter is 3475 km, but again this can't be determined precisely with their model.]

5. The Moon is very big, but not as huge as Earth. Ask, "How does the Moon's diameter compare with the diameter of the Earth?" [The Moon is very big, but only about one-fourth as big as the Earth.]

6. Report the diameter of the Sun. Ask for their measurements of the diameter of the Sun. Their answers will likely vary, but should be about 1,300,000 km. The Sun is super-huge! Its diameter is more than 100 times as big in diameter as the Earth's.

Unit Goals

Size: Some sky objects are relatively small, and some are huge.

Distance: Some objects are relatively close to Earth and some are very far away.

Distance of sky objects from us affects their apparent size: Large objects appear small when far away.

TEACHER CONSIDERATIONS

TEACHING NOTES

Relative Size Surprise: Some students may be stunned to learn the size of the Sun. Their amazement may be because the Sun *looks* small in the sky. In Session 1.5, there will be an opportunity for in-depth discussion of apparent sizes versus real sizes of sky objects.

Optional size challenge: How many Moons would fit across the diameter of the Sun?

QUESTIONNAIRE CONNECTION

The activities in this session deal directly with the information necessary to correctly answer Question #1 on the Pre-Unit 1 Questionnaire.

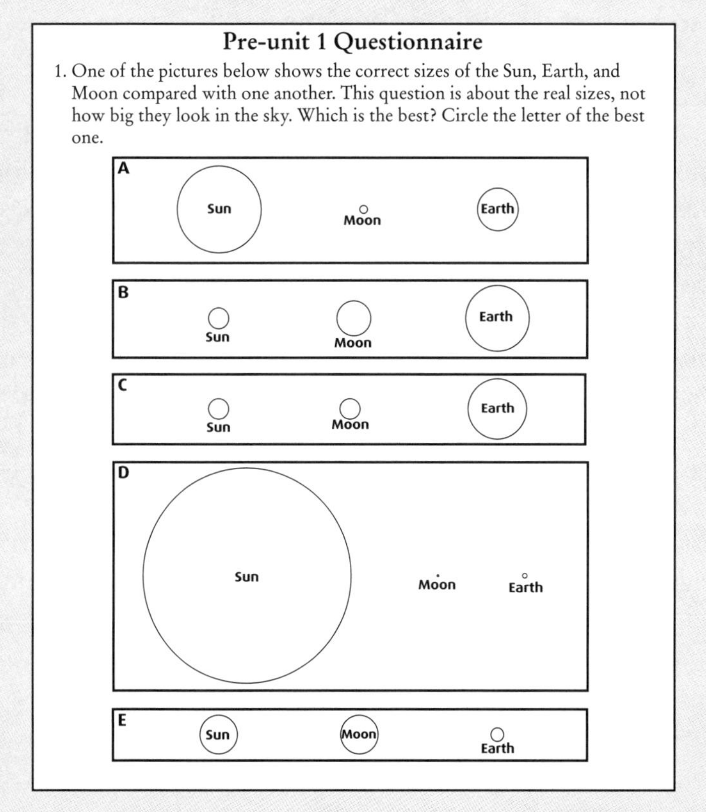

Pre-unit 1 Questionnaire

1. One of the pictures below shows the correct sizes of the Sun, Earth, and Moon compared with one another. This question is about the real sizes, not how big they look in the sky. Which is the best? Circle the letter of the best one.

If the questionnaire results showed that your students had misconceptions about the relative sizes of these objects, you will want to pay close attention to their comprehension during this session. If you notice that your students are still struggling with these concepts, you may choose to spend more time on them.

What One Teacher Said

"They are doing a great job measuring with the rulers. It's amazing to see them trying to be so careful and asking questions when the numbers don't look right to them. They are responding and collecting data as carefully as scientists would."

Key Vocabulary

Science and Inquiry Vocabulary

Evidence
Scientific Explanation
Model
Scale Model
Prediction
Scientist
Three–Dimensional (3-D)
Two–Dimensional (2-D)

Space Science Vocabulary

Atmosphere
Satellite
Orbit
Diameter
Sphere
System

Comparing the Sizes

1. Which is smallest? Ask "Which is smallest of the three?" [The Moon.] Then ask, "Is the Moon smaller than a satellite or an albatross?" [No, it's *much* bigger. Help students remember that the satellites they measured in the last session were only a few meters in diameter. The Moon is very large compared to those sky objects!]

2. Which is biggest? Ask one pair how many Earth diameters they estimate can fit across the Sun's diameter. Say to the rest of the class, "Raise your hand if you got the same estimate." Ask, "Who got more?" "Less?" Focus in on a range that represents most students' estimates. Tell them about *109 Earths can fit across the diameter of the real Sun.* Point out that the scale models give us an idea of the Sun's size, but the real Sun is so big that we can't really picture it. It's super-huge!

3. Introduce Sun as a star. Tell students that the Sun is a star. Even though it is so big compared to Earth, it is medium-sized for a star. Let students know they will be learning more about why the Sun looks about the same size as the Moon in the sky and why stars look so small in the sky when they are really so big.

4. Show the 3-D models. Hold up the card with the 4 mm Earth ball and the 1 mm Moon ball. Walk around so students can see them and say that it is important to remember that the real Earth, Moon, and Sun are round like balls, not two-dimensional. Then reveal the big 3-D Sun. Say that the size differences are even more striking with 3-D models!

5. Order objects from smallest to largest. Direct their attention to the size challenge at the bottom of their sheets. Ask what order they put the objects in. Use their comments to make a list on the board, starting with the smallest, and discuss any placements they disagree on until you arrive at an order like this: smallest bird, *Sputnik* satellite, largest bird, mountain, Moon, Earth, Sun.

Unit Goals

Size: Some sky objects are relatively small, and some are huge.

Distance: Some objects are relatively close to Earth and some are very far away.

Distance of sky objects from us affects their apparent size: Large objects appear small when far away.

TEACHER CONSIDERATIONS

TEACHING NOTES

Observe: Circulate and observe the strategies students use to find out how many Earth diameters fit across the diameter of the Sun. Some students may make marks about the size of the Earth all across the Sun and count them. Others might measure carefully to the halfway point then stop and multiply by two. Some may use the ruler to measure the Earth's and the Sun's diameters, and divide.

The Models: Place the 3-D models of the Sun, Earth, and Moon on a wall where students can see them up close over the coming days. These models will be used again in Session 1.8 when students measure distances.

SCIENCE NOTES

Comparing the volume of the Sun and Earth: The volume of the Sun is 1.3 million times that of Earth. The Sun contains over 99% of the mass in the Solar system.

What One Teacher Said

"The 3D models helped the students to see the relative sizes of the Sun, Moon and Earth. They were quite astonished."

Key Vocabulary

Science and Inquiry Vocabulary

Evidence
Scientific Explanation
Model
Scale Model
Prediction
Scientist
Three–Dimensional (3-D)
Two–Dimensional (2-D)

Space Science Vocabulary

Atmosphere
Satellite
Orbit
Diameter
Sphere
System

6. Add to Concept Wall. Read the following key concepts aloud, one at a time, and post them on the Concept Wall under *What We Have Learned About Space Science.*

Some objects in the sky, such as the Sun, Moon, stars and planets, are very large.
Other objects in the sky, such as birds, satellites and airplanes, are relatively small.
The Earth is very large.
The Moon is very large, but not as large as the Earth.
The Sun is super huge compared to the Earth.
The Sun is a star. Compared to other stars it is medium sized.

Unit Goals

Size: Some sky objects are relatively small, and some are huge.

Distance: Some objects are relatively close to Earth and some are very far away.

Distance of sky objects from us affects their apparent size: Large objects appear small when far away.

TEACHER CONSIDERATIONS

OPTIONAL PROMPTS FOR WRITING OR DISCUSSION

You may want to choose one or more of the prompts below for science journal writing during class or as homework. Or the prompts could be used for a discussion or during a final student sharing circle in which each student gets a turn to share.

- The model car was an example of a scale model. What are some other scale models you can think of?
- What are some examples of models that are not scale models? [Some of their drawings of sky objects, most teddy bears, many cartoons, certain dolls and toys, and so on.]
- Look at the 3-D models of the Earth, Moon, and Sun: What do you find interesting about the sizes of the Earth, Moon, and Sun compared with one another?
- Can you think of any objects in space that might be
  —smaller than the Moon?
  —bigger than Earth?
  —bigger than the Sun?

ONE TEACHER SUGGESTED

Read to the class: *Actual Size* by Steve Jenkins. Houghton Mifflin Company, 2004.

Key Vocabulary

Science and Inquiry Vocabulary

Evidence

Scientific Explanation

Scale Model

Prediction

Scientist

Three Dimensional (3-D)

Two Dimensional (2-D)

Space Science Vocabulary

Atmosphere

Satellite

Orbit

Diameter

Sphere

System

SESSION 1.5
Sizes Near and Far

Overview

Having gathered evidence about the actual sizes of the Earth, Moon, and Sun in Session 1.4, students now need to reconcile this information with their own perceptions of the apparent sizes of these objects in the sky. In this session, students learn that objects in the sky may appear drastically different in size than they actually are because of their different distances from Earth.

The class measures a piece of paper from up close and from across the room, then discusses why the measurements are different. The same procedure is followed in measuring a student's height from near and far. The students are given a picture including various objects in the sky and on the ground, and challenged to rank the objects by size. The objects have been chosen and drawn to provoke discussion about how difficult it is to tell the size of objects when some are near and some are very far.

Students meet in groups of four to discuss their ideas about the actual sizes of the objects and take turns arranging a set of cards to show how they think these objects should be ranked. Students are then given the real measurements of the objects, and have the opportunity to re-order some of the objects based on this evidence.

The session concludes with a short writing assignment that helps students synthesize the concepts introduced and builds their skills in making evidence-based explanations. Evidence-based explanations will be emphasized again during the "evidence circles" activities in Sessions 1.8 and 1.9, as well as in other units of the *Space Science Sequence*.

Sizes Near and Far	Estimated Time
Measuring a paper and a student from near and far	10 minutes
Ranking objects by actual size	20 minutes
Introducing new evidence and re-sorting the objects	10 minutes
Key concepts about size and distance	10 minutes
Writing assignment	10 minutes
TOTAL	**60 minutes**

Unit Goals

Size: Some sky objects are relatively small, and some are huge.

Distance: Some objects are relatively close to Earth and some are very far away.

Distance of sky objects from us affects their apparent size: Large objects appear small when far away.

TEACHER CONSIDERATIONS

Key Vocabulary

Science and Inquiry Vocabulary

Evidence

Scientific Explanation

Model

Scale Model

Prediction

Scientist

Three–Dimensional (3-D)

Two–Dimensional (2-D)

Space Science Vocabulary

Atmosphere

Satellite

Orbit

Diameter

Sphere

System

What You Need

For the class

- ❑ overhead projector or computer with large screen monitor/LCD projector
- ❑ overhead transparency of *A Ship, a Boat, Two Hills, and a Cloud,* from the transparency packet, or CD-ROM file
- ❑ scissors or a paper cutter to cut up cards
- ❑ a few paper clips
- ❑ a piece of chart paper
- ❑ wide-tip felt pen
- ❑ sentence strips for four key concepts
- ❑ 1 measuring tape

For each group of four students

- ❑ 1 copy of the *Cards for Ten Objects* sheet, from the student sheet packet

For each pair of students

- ❑ 1 metric measuring tape or ruler

For each student

- ❑ 1 copy of *Ten Objects* sheet from the student sheet packet
- ❑ 1 piece of writing paper

Getting Read

1. Arrange for the appropriate projector format (computer with large screen monitor, LCD projector or overhead projector) to display images to the class.

2. If you will not be using the CD-ROM, make an overhead transparency *of A Ship, a Boat, Two Hills, and a Cloud.*

3. Photocopy one *Cards for 10 Objects* sheet for each group of four students. Cut the 10 cards and clip a set for each group.

4. On a piece of chart paper, make the chart, Actual Sizes of 10 Objects. Don't post it yet, but have it ready to post later in the session.

Unit Goals

Size: Some sky objects are relatively small, and some are huge.

Distance: Some objects are relatively close to Earth and some are very far away.

Distance of sky objects from us affects their apparent size: Large objects appear small when far away.

TEACHER CONSIDERATIONS

TEN OBJECTS	ACTUAL SIZES
House	25 meters long
Person A	1 meter tall
Person B	1 meter 70 cm. tall
Moon	3500 km diameter
Ball	60 cm diameter
Star A: Betelgeuse	900,000,000 km diameter
Star B: Sirius B	12,000 km diameter
Star C: Rigel	84,000,000 km diameter
Venus	12,100 km diameter
Car	25 cm long

TEACHING NOTES

It's a toy car!: Note that the actual size of the car is only 25 cm long. This is deliberate.

Key Vocabulary

Science and Inquiry Vocabulary

Evidence
Scientific Explanation
Model
Scale Model
Prediction
Scientist
Three–Dimensional (3-D)
Two–Dimensional (2-D)

Space Science Vocabulary

Atmosphere
Satellite
Orbit
Diameter
Sphere
System

5. Write the following four key concepts on sentence strips and have them ready to post on the concept wall:

How big something looks and how big it really is can be very different.
An object looks bigger when it's closer. An object looks smaller when it's farther away.
The Sun looks bigger than other stars because it's a whole lot closer.
The Sun looks the same size as the Moon because it's much farther away than the Moon.

Measuring an Object Near and Far

1. Review the relative sizes of Earth, Moon, Sun. Ask, "From your measurements of the models in the last session, how big is the Earth compared to the Sun?" [Over one hundred Earths could fit in a line across the diameter of the Sun.] "How does the Moon compare in size to the Sun?" [The Moon is only about one-fourth the size of the Earth in diameter, so the Sun is much bigger than the Moon.] Assure your students that scientists have much evidence that these are the real sizes of the Earth, Moon, and Sun.

2. Ask why the Sun and the Moon look about the same size. Ask, "If the Sun is really so huge compared to the Moon, why do they look about the same size in the sky?" [Accept all answers for now.]

3. Measure a piece of paper. Say that objects can look smaller or larger depending on how far away they are. Hand out a measuring tape or ruler and any sheet of 8.5 x 11 inch paper to each pair of students and ask them to measure how long the sheet is, top to bottom, in centimeters. [28 cm.]

4. Measure the same paper from across the room. Now tape a sheet of the same size paper on the wall. Tell your students to stay in their seats, and use their tapes or rulers to measure it from a distance. Tell them it may help to close one eye.

Unit Goals

Size: Some sky objects are relatively small, and some are huge.

Distance: Some objects are relatively close to Earth and some are very far away.

Distance of sky objects from us affects their apparent size: Large objects appear small when far away.

TEACHER CONSIDERATIONS

What One Teacher Said

"The students were intrigued with the difference in measurements when measuring the paper in front of them, and then again across the room."

Key Vocabulary

Science and Inquiry Vocabulary

Evidence

Scientific Explanation

Model

Scale Model

Prediction

Scientist

Three–Dimensional (3-D)

Two–Dimensional (2-D)

Space Science Vocabulary

Atmosphere

Satellite

Orbit

Diameter

Sphere

System

5. Report measurements. When they have finished measuring, tell them to raise their hands when you announce the number of centimeters that matches their measurement. Say "1 centimeter, 2 centimeters," etc., briefly pausing after each one to give them time to raise their hands.

6. Ask why the measurements don't agree. Point out that just a few minutes ago they agreed the paper was 28 centimeters long. Ask how they could be coming up with smaller measurements now. Did the paper change size? [Your students will probably tell you that things look bigger when they're closer and smaller when they're far away. Some students are closer to the paper than others, so their measurements are larger.]

Measuring a Student From Near and Far

1. Select a volunteer. Ask for a volunteer to be measured. (Try not to choose someone who might be sensitive about his or her height.)

2. Measure and record actual height. Ask two other student volunteers to come forward and measure how tall the student is, then return to their seats. Write this measurement on the board.

3. Have students measure the volunteer from their seats. Next, have the same student stand on some kind of elevated platform (such as a chair or table) so all students can see him/her. Pass out a measuring tape to each pair of students and ask them to measure this student from their seats. Make sure they know to measure the student (in centimeters) from foot to head—not from the floor.

4. Report measurements. Once again, when they have finished measuring, begin announcing centimeters and ask them to raise their hands when you say the number of centimeters that matches their measurement.

5. Ask, "Why don't the measurements agree?" Record the range of measurements on the board next to the earlier measurement. Again ask how they could be coming up with different measurements, when the student has not changed size.

Unit Goals

Size: Some sky objects are relatively small, and some are huge.

Distance: Some objects are relatively close to Earth and some are very far away.

Distance of sky objects from us affects their apparent size: Large objects appear small when far away.

Key Vocabulary

Science and Inquiry Vocabulary

Evidence

Scientific Explanation

Model

Scale Model

Prediction

Scientist

Three–Dimensional (3-D)

Two–Dimensional (2-D)

Space Science Vocabulary

Atmosphere

Satellite

Orbit

Diameter

Sphere

System

OBJECTS FROM BIGGEST TO SMALLEST CARDS

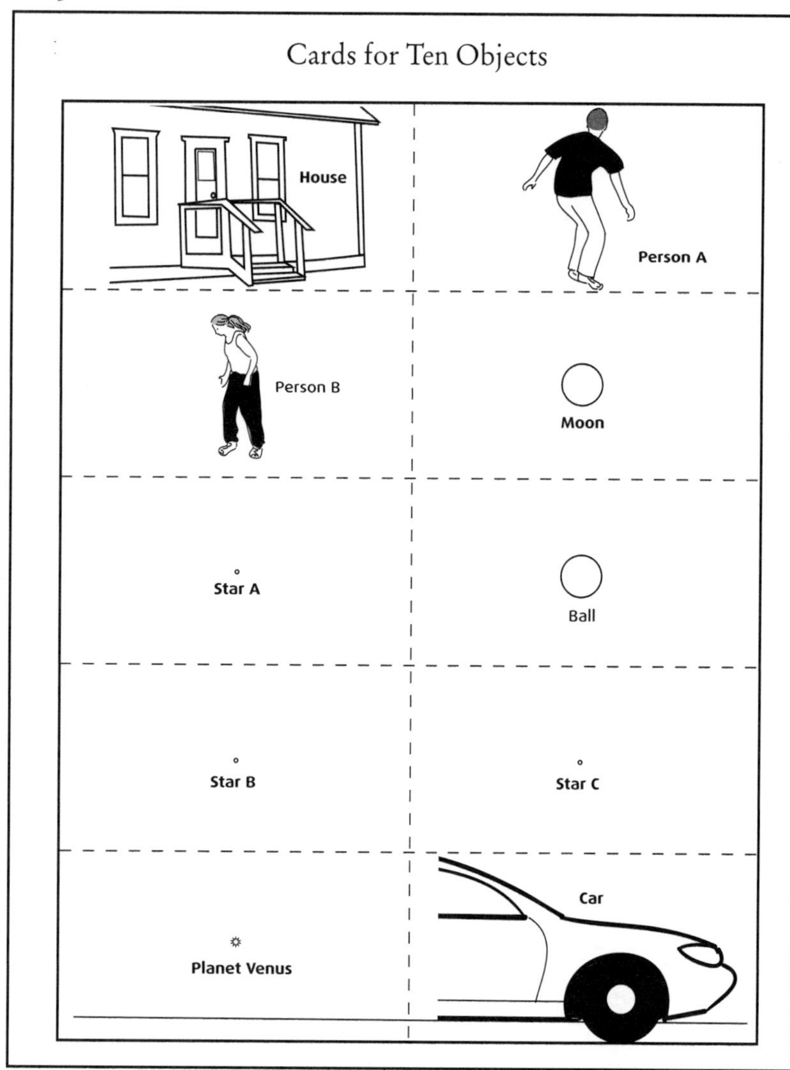

6. Compare actual height with perceived height. Tell them that the first measurement is how tall the student IS, and the other measurements are how tall the student LOOKS or appears when he or she is farther away. Emphasize that the real size of the student has not changed. Collect the measuring tapes.

Show How to Rank Objects by Actual Size

1. Show transparency. Show *A Ship, a Boat, Two Hills, and a Cloud* to the class.

2. Demonstrate how to discuss and rank objects by size. Ask them to list the objects in the picture from biggest to smallest, by the size they **really are**, not the sizes they look in the picture. Ask a few questions about the relative sizes, for example, "Is the ship really bigger than the cloud?" [It looks bigger, but maybe the cloud is far away and really bigger than the ship.] Even if the class doesn't come to agreement, demonstrate how to write a #1 near what they think is the largest object, #2 near the next largest, and so on.

3. Picture of ten objects. Tell students they will each get a sheet with pictures of ten objects. They will try to put them in order by the size they think each object really is, in the same way you just demonstrated. They will write a number near each object, with #1 being the biggest, and #10 being the smallest. Acknowledge that it may be difficult to be sure, but they should take their best guess.

4. Pass out the Ten Objects sheet. Give students a few minutes to work independently to number the objects from 1 to 10.

TEACHER CONSIDERATIONS

OBJECTS FROM BIGGEST TO SMALLEST

A SHIP, A BOAT, TWO HILLS, AND A CLOUD

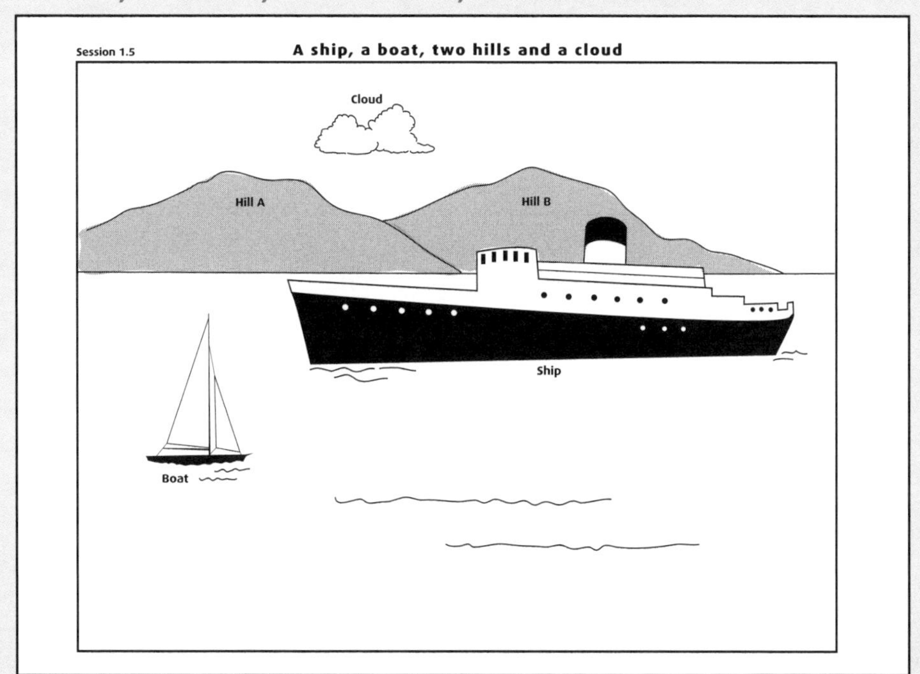

TEN OBJECTS

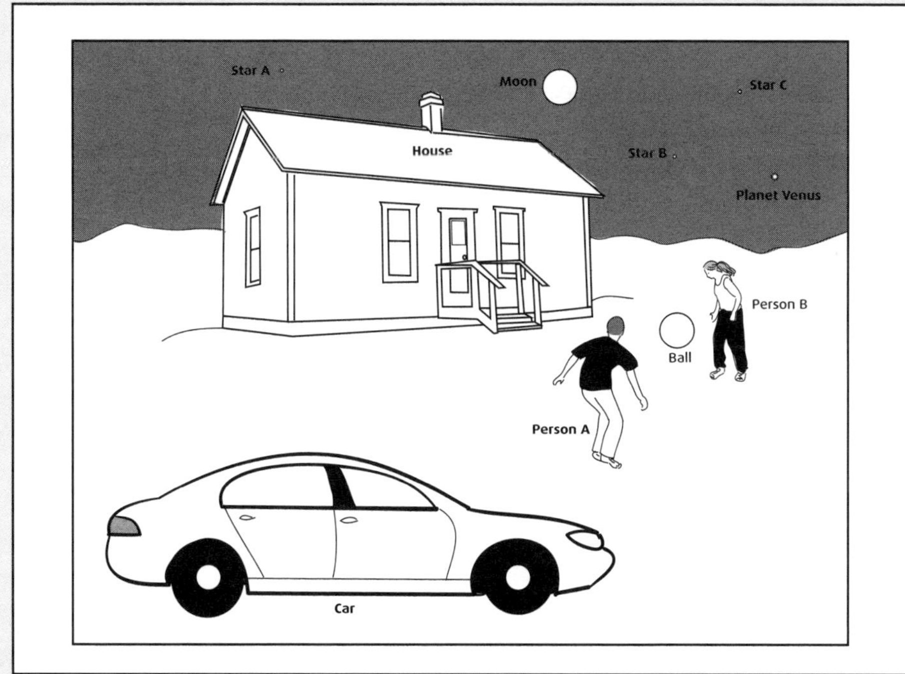

Working in Groups to Discuss and Rank the Objects

1. Explain the cards. Regain the attention of the class. Tell them they will now discuss the sizes of the objects with their group of four. Each group will receive a set of 10 cards to help them during the discussion. Each card shows one of the objects in the picture. Each group will try to agree how to put the cards in order from biggest to smallest.

2. Explain the sorting and discussion procedure. Tell them each person in their group will take a turn putting the cards in the order they think is best, and explaining why. Emphasize that *explaining why* is an important part of the task.

3. Clarify the procedure. Ask, "Should you get into loud arguments or make fun of what others say?" [No.] Ask, "How might you disagree while being respectful?" Emphasize the importance of being polite, and go over the procedure:

a. When it is someone's turn, that person gets to put the cards in whatever order they think is most accurate and explain why they think they belong in that order.

b. When that person has finished, anyone who ***agrees*** with all or part of it can say so and say why.

c. Then anyone who ***disagrees*** with all or part of it can say what they disagree with and why.

d. After each person has a turn to sort the cards and explain, the group will discuss, and try to agree.

4. Distribute cards. Pass out a set of 10 cards to each group and have them follow the discussion procedure. Circulate to make sure students are explaining their reasons, and to check that they are taking turns. If needed, give groups a five-minute warning before the end of the activity so they can make sure everyone has a turn.

Introducing New Evidence and Re-Sorting

1. Review the meaning of *evidence*. Have students leave their cards on their tables, and regain the attention of the class. Refer to the definition of evidence on the concept wall. Tell them that in the last session, they gathered evidence about the real size of the Moon. Ask what that was. [3,000 km diameter.]

Unit Goals

Size: Some sky objects are relatively small, and some are huge.

Distance: Some objects are relatively close to Earth and some are very far away.

Distance of sky objects from us affects their apparent size: Large objects appear small when far away.

TEACHER CONSIDERATIONS

TEACHING NOTES

How do scientists gather evidence about stars? Students may be curious to know how scientists gather evidence about faraway objects. Astronomers use measurements of starlight color and brightness as well as mental models to determine qualities of stars, such as how big they are, how hot they are, and what they are made of. For your own interest, the *Background* section (page 25) contains more detailed information about measuring the sizes and distances of stars. However, these concepts are generally too advanced for elementary school students. For now, assure them that scientists have been able to gather evidence about the sizes and distances of many stars. Encourage students to learn more about this in future years.

Key Vocabulary

Science and Inquiry Vocabulary

Evidence

Scientific Explanation

Model

Scale Model

Prediction

Scientist

Three–Dimensional (3-D)

Two–Dimensional (2-D)

Space Science Vocabulary

Atmosphere

Satellite

Orbit

Diameter

Sphere

System

2. Point out the shortage of evidence. Admit that they haven't been given evidence about the real sizes of the other nine objects in the picture. They have done the best they could, but it is sometimes difficult to tell the real sizes of objects just by looking at a picture.

3. Scientists need evidence. Tell students that many great thinkers in ancient times struggled to understand the real sizes of objects in the sky, but they were often very far off in their estimates. They had limited evidence to work with. When space scientists were able to find better ways to measure far-away objects, people were able to correct their models.

4. Scientists change their minds based on new evidence. Scientists think about and discuss ideas and evidence in order to find the best explanation for something—an explanation that best matches all available evidence. Say one of the signs of a true scientist is the ability to listen to others and change your mind when you find that what you think doesn't match the evidence.

5. Introduce new evidence about the objects. Tell them you're going to give them more evidence to work with—the real measurements of the objects in the picture. Put the list you prepared earlier on the wall.

TEN OBJECTS	ACTUAL SIZES
House	25 meters long
Person A	1 meter tall
Person B	1 meter 70 cm tall
Moon	3500 km diameter
Ball	60 cm diameter
Star A: Betelgeuse	900,000,000 km diameter
Star B: Sirius B	12,000 km diameter
Star C: Rigel	84,000,000 km diameter
Venus	12,100 km diameter
Car	25 cm long

6. Briefly discuss surprises. Students may be surprised to find out that:

- The car is only 25 cm long—it must be a toy car in the foreground.
- Person B is actually taller than Person A.
- The stars are all much bigger than anything else in the picture.

Unit Goals

Size: Some sky objects are relatively small, and some are huge.

Distance: Some objects are relatively close to Earth and some are very far away.

Distance of sky objects from us affects their apparent size: Large objects appear small when far away.

TEACHER CONSIDERATIONS

ASSESSMENT OPPORTUNITY AND QUESTIONNAIRE CONNECTION

Critical Juncture—Size and Distance

2. Why does the Sun look much bigger than the stars we see at night? Circle the letter of the best answer.
   A. The Sun is much bigger than the stars.
   B. The Sun looks bigger because it is closer to us than the stars.
   C. The Sun looks bigger because it is farther away from us than the stars.
   D. The Sun can be seen only in the daytime.

4. Why do the Sun and Moon look as though they are about the same size in the sky? In your answer, explain how big and how far away they are.

   ____

   ____

The activities in this session deal directly with the information necessary to correctly answer Questions #2 and #4 on the *Pre-Unit 1 Questionnaire.* If you noticed in the questionnaire results that your students tended to struggle with these questions, you will want to pay special attention to their comprehension during this session.

Quick check for understanding: During discussions, if you notice some uncertainty about how distance affects apparent size, you may want to allow time for one or both of the activities in Providing More Experience on page 103.

OPTIONAL PROMPT FOR WRITING OR DISCUSSION

You may want to use the additional prompt below for science journal writing during class, for homework, or for a discussion.

Pretend someone said, "The Moon is much bigger than the stars. I can tell that because it looks so much bigger in the sky." Write a paragraph to explain to the person why the Moon looks bigger than the stars. Include how big the Moon and stars really are.

Key Vocabulary

Science and Inquiry Vocabulary

Evidence
Scientific Explanation
Model
Scale Model
Prediction
Scientist
Three–Dimensional (3-D)
Two–Dimensional (2-D)

Space Science Vocabulary

Atmosphere
Satellite
Orbit
Diameter
Sphere
System

7. Groups re-sort the cards, and each student writes the list. Have groups re-order their cards, taking the new evidence into account. Say that each student will then write a list of the objects from biggest to smallest at the bottom right-hand area of their sheet (largest object at the top as #1, and smallest at the bottom as #10).

Key Concepts About Size and Distance

1. Discuss the actual sizes of stars. After a few minutes, get the attention of the class. Remind students of the diameter of the Sun measured in the last session (1,300,000 km), and refer to the key concept that the Sun is a medium-sized star.

2. Stars are different sizes. Say that scientists have been able to measure the sizes of many stars, and there are some stars larger than the Sun, some smaller, and some about the same size. Ask, "Why do the three stars in the picture look the same size if each star is a different size?" [The biggest star pictured is much farther away than the others, etc.]

3. Ask, "Why does the Sun look about the same size as the Moon in the sky?" [The Sun is really much bigger than the Moon, but it looks about the same size because it is much farther away from us than the Moon.]

4. Post key concepts. Read aloud and post the following key concepts under *What We Have Learned About Space Science.*

How big something looks and how big it really is can be very different.
An object looks bigger when it's closer. An object looks smaller when it's farther away.
The Sun looks bigger than other stars because it's much closer.
The Sun looks the same size as the Moon because it's farther away than the Moon.

Unit Goals

Size: Some sky objects are relatively small, and some are huge.

Distance: Some objects are relatively close to Earth and some are very far away.

Distance of sky objects from us affects their apparent size: Large objects appear small when far away.

TEACHER CONSIDERATIONS

ASSESSMENT OPPORTUNITY

Embedded Assessment—Why Do the Stars Look Small? The short writing assignment at the end of the session can be used as an assessment of student understanding of the idea that distance affects how big an object looks.

Pretend someone said, "The Sun is much bigger than all the other stars. I can tell, because the Sun looks much bigger than the stars in the sky." Explain to the person why the other stars look smaller than the Sun. Include evidence about how big the Sun and other stars really are.

A rubric is provided below that can be used to assess student understanding. The assessment can also be scored using the general rubrics that are included on page 66.

Understanding Science Concepts

The key science concepts for this assessment are the following:

1. How big something looks and how big it really is can be very different.
2. An object looks bigger when it's closer. An object looks smaller when it's farther away.
3. The Sun looks bigger than other stars because it's much closer.

Unit Goals

1. Distance of sky objects from us affects their apparent size: Large objects appear small when far away.

Score	Description
4	The student demonstrates a complete understanding of all of the science concepts and uses evidence to support the written explanation. The student demonstrates an understanding of the key science concepts. The student uses real measurements of the Sun (from Session 1.4) and the stars (from Session 1.5) as evidence to back up their explanations and to demonstrate/illustrate how apparent size changes with distance.
3	The student demonstrates a partial understanding. The student demonstrates an understanding of the key science concepts. However, the student does not tie all of these concepts together in their explanation and support it with evidence from the class activities with measurements of the Sun and Stars. The evidence they use maybe more experiential or statement of fact than from their scientific classroom experiences.
2	The student demonstrates an insufficient understanding of the science concepts. The student demonstrates an understanding of one or two of the key concepts. However, the student does not demonstrate an understanding of all of the concepts and does not use evidence from class to support their explanation.
1	The content information is inaccurate. Some possible inaccuracies are 1. The Sun is bigger than stars (implying that the Sun is not a star). 2. The Sun is bigger than all the other stars in the sky. 3. The Sun looks bigger in the sky since it is actually bigger than the other stars. 4. The Sun looks bigger in the sky since it is further away than the other stars.
0	The response is irrelevant or off topic.
n/a	The student has no opportunity to respond and has left the question blank

Key Vocabulary

Science and Inquiry Vocabulary

Evidence

Scientific Explanation

Model

Scale Model

Prediction

Scientist

Three–Dimensional (3-D)

Two–Dimensional (2-D)

Space Science Vocabulary

Atmosphere

Satellite

Orbit

Diameter

Sphere

System

Writing an Evidence-Based Explanation

1. Write the following assignment on the board or overhead: Pretend someone said, “The Sun is much bigger than all the other stars. I can tell, because the Sun looks much bigger than the stars in the sky.” Explain to the person why the stars look smaller than the Sun. Include evidence about how big the Sun and stars really are.

2. Pass out paper or journals and have them begin. If time is short, you might have them finish the assignment for homework.

Unit Goals

Size: Some sky objects are relatively small, and some are huge.

Distance: Some objects are relatively close to Earth and some are very far away.

Distance of sky objects from us affects their apparent size: Large objects appear small when far away.

TEACHER CONSIDERATIONS

PROVIDING MORE EXPERIENCE

1. Soccer ball and baseball. If your students are struggling with the idea of how distance affects perceived size, you may choose to use this example to provoke further discussion. Project an overhead transparency of the example below, and discuss the question.

This is a picture of a soccer ball and a baseball in the air. Why do they look as though they are the same size?

2. Comparing Two Figures, Near and Far

a. Photocopy a drawing or picture of an object, so that you have two identical copies. Post one on the wall where students can see it. Hold the second copy somewhere near the center of your students, away from the drawing posted on the wall. Ask students to raise their hands if

- they think the picture on the wall is bigger than the one you're holding
- they think the picture you're holding is bigger than the one on the wall
- they think the pictures are the same size

b. Now, tape the picture in your hand on the wall, and take the one that was posted on the wall to the center of your students. Ask the same questions again.

c. Post the two pictures on the wall next to each other. Measure exactly how tall each one is, and announce these measurements to the class.

d. Ask, "Why did it seem that one was bigger than the other when they were not next to each other?"

e. Tell them that if it's hard to tell the size of objects this close, then they can understand why it's hard to tell the size of things in the sky. When you look at something as far away as a star, it's hard to know how big it really is.

Key Vocabulary

Science and Inquiry Vocabulary

Evidence

Scientific Explanation

Model

Scale Model

Prediction

Scientist

Three–Dimensional (3-D)

Two–Dimensional (2-D)

Space Science Vocabulary

Atmosphere

Satellite

Orbit

Diameter

Sphere

System

SESSION 1.6

Ranking Space Objects by Size

Overview

Although the primary focus of Unit 1 is on the Earth, Moon, and Sun, this session temporarily digresses to explore other space objects that your students may be curious about. The focus is still on size, allowing students to put space objects into context based on the sizes of the Earth, Moon, and Sun. They learn that there are many space objects out there that are inconceivably large. The session also helps deepen their understanding that huge space objects may look small because they are so far away.

Each team of two receives a set of 27 cards with images of sky objects on them. Teams begin by examining and discussing each one. Next, they categorize the cards by size, while comparing each sky object to the size of a school, the Moon, the Earth, and the Sun.

When they have completed this task, the teacher leads the class in a *Tour of Sky Objects*. Images and information about the sizes of the objects, as determined by space scientists, are revealed, and students adjust their card sorts accordingly. The tour serves the dual purpose of revealing the actual relative sizes of the objects and providing an opportunity for students to gaze in wonderment at beautiful and mysterious space images. Although student questions and discussion will likely lead in many directions, the focus is maintained on the *relative sizes* of these objects.

Ranking Space Objects by Size	Estimated Time
Exploring *Sky Object Cards*	5 minutes
Categorizing *Sky Object Cards*	20 minutes
Tour of Sky Objects	30 minutes
The *Sizes of Sky Objects* chart	5 minutes
TOTAL	**60 minutes**

What You Need

For the class

- ❑ overhead projector or computer with large screen monitor/LCD projector
- ❑ overhead transparency from the transparency packet or CD-ROM file of completed *Sizes of Sky Objects* chart
- ❑ set of 27 images for the *Tour of Sky Objects*; these can be CD-ROM files or transparencies from the transparency packet
- ❑ wide-tip felt pen

Unit Goals

Size: Some sky objects are relatively small, and some are huge.

Distance: Some objects are relatively close to Earth and some are very far away.

Distance of sky objects from us affects their apparent size: Large objects appear small when far away.

TEACHER CONSIDERATIONS

CD-ROM NOTES

Tour of Sky Objects: If you are using the CD-ROM version to present the tour to the class, practice using it ahead of time. Open the *Tour of Sky Objects* program. Click the "back" or "next" buttons on the screen with the mouse to scroll through a photo album of sky objects arranged in groups from smallest to largest. On the bottom of the screen there is a progress bar showing you where, on a scale from bird to Universe, the sky object you are looking at falls. You can also jump ahead or back on the bar by clicking on the "signpost" icons, such as school or Sun. Further instructions for using this program are included on the CD-ROM.

Another resource on the CD-ROM for independent student exploration: On the CD-ROM, you will also find a student-centered activity called *What's Up?* This is not designed to be used with the whole class, but rather as an enrichment opportunity if you have a computer available for students to use individually. In *What's Up?* students can click on a wide variety of sky objects to find information about the objects. Students can explore objects they find in the night and day sky using this fun, interactive resource. By clicking on objects in the sky with their mouse (such as *bird, Sun,* or *parachute* in the day sky, or *satellite, Moon,* or *rocket* in the night sky) students will learn the sizes of these objects and their distances from the surface of the Earth as the information appears on the bottom of the screen. Students also have the option of clicking on the "click here for more images" on the small photo in the lower left-hand corner; this opens an interactive photo album of the sky object they just selected. Students control the photo album by clicking "back" and "next" with the mouse, and "close" when they are finished. Informational text concerning the class of sky object selected, or the specific photo, accompanies each image. Students can switch back and forth between day and night by clicking the "day" or "night" buttons in the lower center of the screen with the mouse. Further instructions for using this program are included on the CD-ROM.

Key Vocabulary

Science and Inquiry Vocabulary

Evidence

Scientific Explanation

Model

Scale Model

Prediction

Scientist

Three Dimensional (3-D)

Two Dimensional (2-D)

Space Science Vocabulary

Atmosphere

Satellite

Orbit

Diameter

Sphere

System

- ❑ 1 sentence strip for the key concept
- ❑ *optional:* student drawings of sky objects from Session 1.2

For each team of two

- ❑ set of 27 *Sky Object Cards* and 5 *Category Cards*, from the student sheet packet
- ❑ 1 envelope or sandwich bag to hold a set of cards
- ❑ 1 copy of *Visual Chart of Sizes of Planets in the Solar System* from the student sheet packet
- ❑ approximately 10 small sticky notes

Getting Ready

1. Arrange for the appropriate projector format (computer with large screen monitor, LCD projector, or overhead projector) to display images to the class.

2. If you will not be using the CD-ROM, make the following overhead transparencies:
   - *Sizes of Sky Objects* chart
   - set of 27 transparencies for the *Tour of Sky Objects*

3. Make enough copies of the *Sky Object Cards* and *Category Cards* sheets to be sure that you have one set per team of two students.

4. Cut apart the 27 sky object cards and five category cards. Make one set for each team of two students, and put each set in a separate envelope. *Optional: Laminate the cards before cutting.*

5. Make stacks of about 12 small sticky notes for each team of two.

6. Write the following key concept on a sentence strip and have it ready:

There are many things in the Universe that are much larger than the Sun.

Exploring Sky Object Cards

1. What are the sizes of other sky objects compared with the Earth, Moon, and Sun? Remind your students that during the previous sessions they compared the sizes of the Earth, Moon, and Sun, and learned that the Moon is very big, the Earth is huge, and the Sun is super huge. Tell them that in this session, they will be comparing the sizes of other space objects with these three.

TEACHER CONSIDERATIONS

Key Vocabulary

Science and Inquiry Vocabulary

Evidence

Scientific Explanation

Model

Scale Model

Prediction

Scientist

Three–Dimensional (3-D)

Two–Dimensional (2-D)

Space Science Vocabulary

Atmosphere

Satellite

Orbit

Diameter

Sphere

System

2. Examine the Sky Object Cards. Say that each pair of students will receive a set of cards with pictures of sky objects on them. They will look at each picture and discuss how big they think each sky object is.

3. Teams of two receive cards. Divide your students into teams of two. Give each pair a set of cards, and let them observe and discuss the sky objects.

Categorizing Sky Object Cards

1. Introduce five categories for the Sky Object Cards. After allowing a few minutes for exploration of the cards, regain the attention of the class. Tell them that their next job will be to sort the sky objects into five categories by the size of the object pictured. Emphasize that they will sort by the size they think the object *really* is, not the size it appears to be when viewed from Earth. Hold up each size category card, and clarify what it means:

a. **Smaller than a school.** Ask for an example of a sky object that could go in this category. [Just take one example for now, but birds, satellites, spaceships, some clouds, some asteroids, and some comets are all possible answers.] Tell them that if they think a sky object is smaller than a school, they will place that card near this category card on their desk.

b. **Bigger than a school, but smaller than the Moon.** Hold up the second category card. Ask if they can remember the size of the Moon from their earlier measurement of a Moon model. [3,000–km diameter.]

c. **Bigger than the Moon, but smaller than Earth.** Ask them to remember how big the Earth is. [The Earth is four times the size of the Moon—13,000–km diameter.]

d. **Bigger than Earth, but smaller than the Sun.** Ask them to describe the size of the Sun compared to Earth. [109 Earth diameters could fit in a line across the diameter of the Sun.]

e. **Bigger than the Sun.**

Unit Goals

Size: Some sky objects are relatively small, and some are huge.

Distance: Some objects are relatively close to Earth and some are very far away.

Distance of sky objects from us affects their apparent size: Large objects appear small when far away.

TEACHER CONSIDERATIONS

What Some Teachers Said

"One student said, 'You had my mind rumbling!' Comparing sky objects to objects they know about really puts things in perspective."

"The students enjoyed putting the pieces in the right categories. They even enjoyed debating one another as to the placement of several objects. I was impressed by their level of recall on the previous lessons regarding size comparisons of the Earth, Moon, and Sun. This is information I know they will remember."

"Students were well immersed in the lesson. They carried on interesting discussions about what objects go under each of the categories. Students were able to retain past information and use it in this lesson."

2. Sort on desktops. Suggest that they spread the five category cards out across their desks in order from small to big, discuss with their partner where each sky object card should go, and place it in a column under a category card. Tell them that there will probably be many objects that they aren't sure about; they should discuss those and take their best guess. If partners can't agree which category an object goes in, they should put it aside until the class discussion.

3. "Sticky notes" to put objects in more than one category. Say that an object could belong in different size categories. For example, ask, "Where would you put the *asteroid* card?" [An asteroid could be smaller or bigger than a school.] To put *asteroid* in more than one category, they can label a sticky note *asteroid*, and then put the card in one category and the sticky note in the other.

4. Students sort for about 15 minutes. Pass out a set of category cards and a stack of about 12 sticky notes to each team of two.

5. Stop sorting. After about 15 minutes, tell them to stop sorting, whether or not they have finished. Have them leave their sorts on their desks.

Tour of Sky Objects

1. Introduce the tour. Get the attention of the class. Tell them that they may not be sure if they put some of the cards in the right categories. Say that this is okay, and that one of the reasons for the sort was to give them a chance to think about the sizes of sky objects.

2. Adjust their charts. Tell them that space scientists have studied and measured these sky objects with many different techniques and equipment. Let them know that you're now going to show pictures of each of the sky objects and give them the best available information that space scientists have about the real sizes of the objects. Tell them that as you do this, they may adjust the sorts at their tables to match this information.

3. Show the tour. Show the *Tour of Sky Objects* on the overhead projector, or use the CD-ROM version with a large screen monitor or LCD projector. Use the script (pages 178-184) to assist in narration and discussion. Whenever possible, review why huge things look much smaller than they really are when observed from far away.

4. Smallest to largest. Let them know that you'll be starting with the objects that are smaller than a school, and ending with the objects that are bigger than the Sun.

Unit Goals

Size: Some sky objects are relatively small, and some are huge.

Distance: Some objects are relatively close to Earth and some are very far away.

Distance of sky objects from us affects their apparent size: Large objects appear small when far away.

TEACHER CONSIDERATIONS

TEACHING NOTES

How do scientists gather evidence about sizes of faraway objects? As in Session 1.5, students may be curious to know how scientists gather evidence about such faraway objects. Because these techniques are generally too advanced to explain to elementary school students, just present the evidence now, but encourage interested students to learn more in future years. (The *Background* section, page 25, contains some information about measuring sizes and distances of space objects.)

Keep the Focus on Size: During the "tour," it is natural for many discussion topics to come up. These topics are all worthwhile, but keep the focus on the sizes of space objects. Also, large numbers don't have much meaning for many young students. Instead of numbers, it's best to stay focused on comparative sizes—what object is bigger or smaller than another, and why huge things that are far away can look smaller than small things that are closer.

Key Vocabulary

Science and Inquiry Vocabulary

Evidence

Scientific Explanation

Model

Scale Model

Prediction

Scientist

Three–Dimensional (3-D)

Two–Dimensional (2-D)

Space Science Vocabulary

Atmosphere

Satellite

Orbit

Diameter

Sphere

System

Script for *Tour of Sky Objects*

Smaller than a school

1. **Most meteors (shooting stars):** Not counting the trail of light, meteors can be very small. They are often only millimeters in size. Meteors are sometimes called "shooting stars," but they are not really stars. They are dust or small rocks that fall toward the Earth, and heat the air white-hot as they travel through the atmosphere. They can also be rocks as large as 1 kilometer, although these are extremely rare.

2. **Birds:** You measured models of the smallest and largest: An albatross has about a 3–meter, 63–centimeter wingspan; a hummingbird has about a 10–centimeter wingspan.

3. **Satellites:** Satellites come in different sizes, as you measured in Session 1.3.

4. **Spaceship:** The *space shuttle* is about 37 meters long. This is smaller than most schools.

5. **International Space Station:** You didn't measure the International Space Station, the largest human-made satellite. It is 94 meters long. It is smaller than some schools, but bigger than others. (Putting it in both categories with a sticky note would be fine.)

6. **Small clouds:** Clouds can be very tiny. Some are smaller than a school, some larger.

7. **Some asteroids:** Asteroids are small, rocky objects orbiting the Sun. More than 90,000 have been discovered. Most are found between Jupiter and Mars. They usually have an irregular shape. Asteroids discovered so far range in size from 940 kilometers in diameter (the asteroid Ceres) to just 10 meters in diameter. (Some are smaller than a school, some are larger. But none are larger than the Moon.)

8. **Some comets (not counting the tail):** Comets can be many sizes, but they are usually less than 10 km wide. Comet tails can be very long. They can be longer than the Sun's diameter. This means that comets could go into all five categories.

Unit Goals

Size: Some sky objects are relatively small, and some are huge.

Distance: Some objects are relatively close to Earth and some are very far away.

Distance of sky objects from us affects their apparent size: Large objects appear small when far away.

TEACHER CONSIDERATIONS

TEACHING NOTES

Size Disputes: Although the Earth, Moon, and Sun have definite diameters for comparison, schools can be many different sizes. Acknowledging this may prevent some disputes about what category certain objects belong in.

For background information on the objects shown in the *Tour of Sky Objects*, see pages 47-51 of the *Background for Teachers* section.

About meteors/meteorites/meteoroids and "shooting stars."
At this stage, it is probably not important that students understand all the distinctions below. What is important is that they understand that meteors (sometimes called shooting stars) are ***not*** stars.

The terms *meteors, meteorites*, and *meteoroids* each have different meanings:

- A meteoroid is a piece of material drifting in space.
- A meteor is a meteoroid that streaks into Earth's atmosphere, leaving a trail of light as it heats up and glows white-hot. This is what many people call a shooting star.
- A meteorite is what the object is called if it does reach Earth's surface. Most meteoroids are so small that they completely vaporize and never reach Earth's surface.

What Some Teachers Said

"This session really helped students put sizes of sky objects in perspective. They had a lot of misconceptions about the real size of objects. The script was a great help to me and to students."

"All of my students absolutely loved the CD. I had them investigate the What's Up and the Sky Objects. They had a lot of fun and were very engaged in categorizing the sizes of sky objects. And of course the big discussion about Black Holes came up. A lot of good debate on this topic!"

Key Vocabulary

Science and Inquiry Vocabulary

Evidence

Scientific Explanation

Model

Scale Model

Prediction

Scientist

Three–Dimensional (3-D)

Two–Dimensional (2-D)

Space Science Vocabulary

Atmosphere

Satellite

Orbit

Diameter

Sphere

System

Bigger than a school, but smaller than the Moon

9. **Spaceship:** Apollo 11 was launched on the Saturn V rocket, the biggest rocket ever sent into space. The rocket was 111 meters (or 36 stories tall, or three football fields) tall.

   (Some Meteors): It is rare for a meteor to be larger than a school, unless you count the length of the streak in the sky.

10. **Some black holes**: Scientists think black holes can be any size. In a black hole, matter is pressed tightly together. The gravity near a black hole is so strong that even light cannot escape. You can't really see a black hole. The size of black holes varies according to the size of what formed them. When giant stars (4 to 15 times the mass of the Sun) die, they can become black holes that would be larger than a school, but smaller than the Moon. There is also evidence of black holes formed from millions of stars all squashed together. Those black holes could be larger than the Sun (much larger). Black holes of other sizes theoretically can exist, but scientists do not yet have evidence of their existence.

    (Most comets, not counting the tail)

    (Some asteroids)

11. **Large clouds:** Clouds can be very big. This is a photo taken from a satellite high above the Earth. A huge cloud can be more than 1,000 km across.

**Visual Chart of Sizes of Planets in the Solar System

At this point, pass out to each pair of students the Visual Chart of Sizes of Planets in the Solar System, which shows the Sun and sizes of the planets within. Say, "Earth is one of eight planets in the Solar System. All eight planets orbit the Sun. Some of the planets have moons."

12. **Moon of Mars–Phobos:** Phobos is a moon of the planet Mars. It is much smaller than Earth's Moon, but bigger than a school. Most moons in the Solar System are smaller than the Moon that orbits Earth.

13. **"Dwarf Planet" Pluto:** Until recently, Pluto was called the 9th planet. But in 2006, scientists decided to call Pluto a "dwarf planet." Pluto is smaller than Earth's moon. (See Background page 45 for more on Pluto.)

Unit Goals

Size: Some sky objects are relatively small, and some are huge.

Distance: Some objects are relatively close to Earth and some are very far away.

Distance of sky objects from us affects their apparent size: Large objects appear small when far away.

VISUAL CHART OF SIZES OF PLANETS IN THE SOLAR SYSTEM

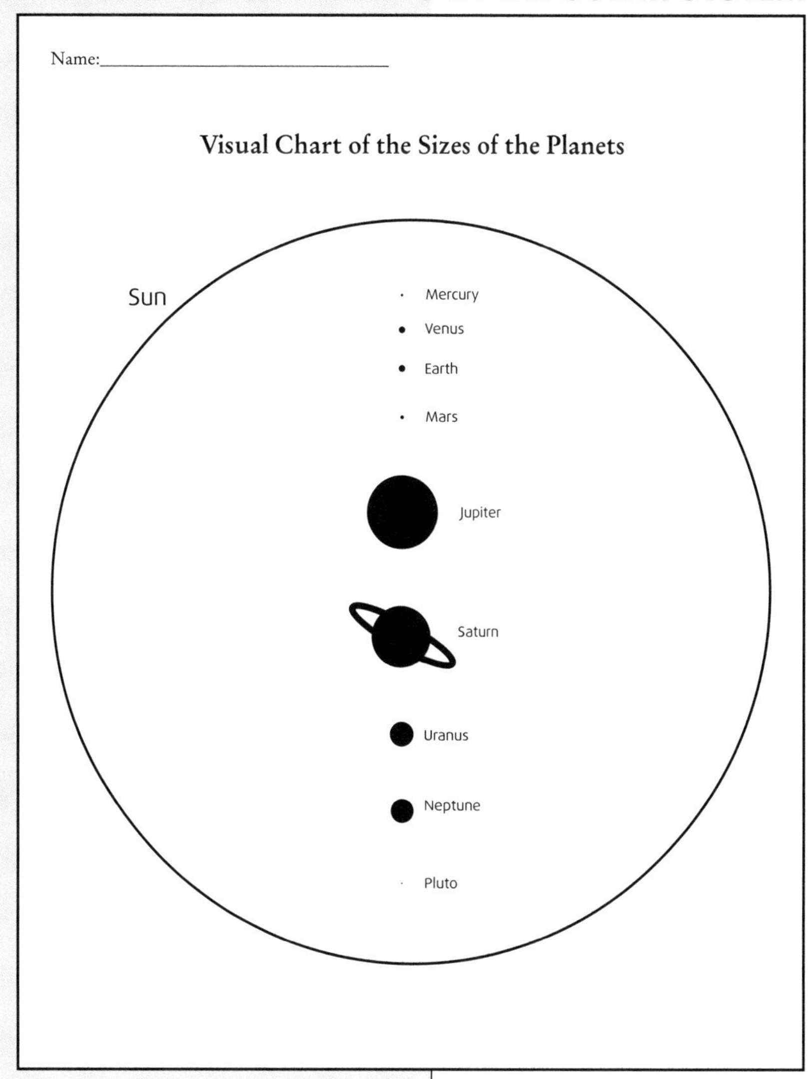

TEACHING NOTES

Extra images on the CD-ROM: The CD-ROM has multiple images of the sky objects that fit into more than one category. The sky objects appear in the same order as they are mentioned in the script. If you are using transparencies, only one image has been provided of each object. You can mention the objects that come up multiple times as you get to them in the script; there is no need to show the image again.

Bigger than the Moon, but smaller than the Earth

(Some Comet Tails)

**Visual Chart of Sizes of Planets in the Solar System

Again, draw their attention to the chart of the planets within the Solar System, and ask the students to use it to list the three planets that are smaller than the Earth, but bigger than the Moon: Mercury, Mars, and Venus.

14. **Planet Mercury:** Mercury is smaller than Earth, and not a whole lot bigger than Earth's Moon. [4,800 km in diameter.]

15. **Moon of Saturn–Titan:** Some moons in the Solar System are larger than the Moon that orbits Earth and larger than some of the smaller planets. Titan orbits the planet Saturn and is 5,150 kilometers in diameter.

16. **Planet Mars:** Mars is smaller than Earth—roughly half the size. [6,400 kilometers in diameter.]

17. **Planet Venus:** Venus is just a bit smaller than Earth. [12,000 km in diameter.] Have any of the students seen Venus in the sky?

Bigger than the Earth, but smaller than the Sun

(Some Comet Tails)

*****Visual Chart of Sizes of Planets in the Solar System***

Again, direct their attention to the chart, and ask students to list the planets that are bigger than the Earth: Neptune, Uranus, Saturn, and Jupiter. (All planets are smaller than the Sun.)

18. **Planet Neptune:** Neptune is about 50,000 km in diameter.

19. **Planet Uranus:** Uranus is about 51,000 km in diameter.

20. **Planet Saturn**: Saturn is about 120,000 km in diameter.

21. **Planet Jupiter:** Four planets in the Solar System are bigger than Earth. Jupiter is the biggest. It is much bigger than Earth (about 140,000 km in diameter).

Unit Goals

Size: Some sky objects are relatively small, and some are huge.

Distance: Some objects are relatively close to Earth and some are very far away.

Distance of sky objects from us affects their apparent size: Large objects appear small when far away.

TEACHER CONSIDERATIONS

Key Vocabulary

Science and Inquiry Vocabulary

Evidence

Scientific Explanation

Model

Scale Model

Prediction

Scientist

Three–Dimensional (3-D)

Two–Dimensional (2-D)

Space Science Vocabulary

Atmosphere

Satellite

Orbit

Diameter

Sphere

System

22. **Some stars:** Our Sun is a medium-sized star. Stars can be much bigger than the Sun, much smaller, or the same size. Every star you can see in the sky is a distant "sun." Why does our star, the Sun, look so much bigger than the others? [Because it is much closer.]

Bigger than the Sun

(Some stars)

(Some comet tails)

23. **Solar System:** The Solar System is made up of the Sun, the planets, moons, asteroids, and comets. A system is a group of objects that form a whole and move or work together. Sol is another name for the Sun, which is why it's called the Solar System. The Sun is the biggest thing in the Solar System by far, but the Solar System is much bigger than the Sun itself. (It is about 60 trillion km in diameter, when measuring out to the Oort Cloud, far beyond Pluto.) This picture shows the planets in the Solar System, but doesn't show the way it really looks, because we can't take a picture of the Solar System. (We are inside it.)

24. **Supernova:** A supernova is an explosion of a very large star.

25. **Nebula:** A nebula is a cloud of gas and dust in space. This nebula is made of material from an exploded star.

(Some black holes)

26. **Galaxy:** A galaxy is much bigger than the Sun and much bigger than the Solar System. There are billions of stars in a galaxy. Our Solar System is a part of the Milky Way galaxy. (Our galaxy has 200–400 billion stars, is about 100,000 light years across in diameter, and 3,000 light years "thick." Most galaxies are smaller.)

27. **Portion of the Universe:** The Universe is our name for everything in space. Of course, we don't have a picture of the Universe, but this is a picture of part of it. Astronomers picked a relatively "boring" part of the sky and, using the Hubble telescope, took a photo of a spot "the size of Eisenhower's eye on a dime held at arm's length." This amazing photo is what they got. Most of the bright blobs in the photo are other galaxies!

Unit Goals

Size: Some sky objects are relatively small, and some are huge.

Distance: Some objects are relatively close to Earth and some are very far away.

Distance of sky objects from us affects their apparent size: Large objects appear small when far away.

TEACHER CONSIDERATIONS

TEACHING NOTES

Objects in the Universe: The tour provides only a quick introduction to objects in the Universe and their relative sizes. The scale and arrangement of objects in the Universe are taught for full understanding in the *Space Science Sequence for grades 6–8.*

SIZES OF SKY OBJECTS

1.6 Transparency: Sizes of Sky Objects

Smaller than a school	Bigger than a school, but smaller than the Moon	Bigger than the Moon, but smaller than the Earth	Bigger than the Earth, but smaller than the Sun	Bigger than the Sun
Birds	*Apollo 11* Saturn V Rocket	Planet Mercury	Planet Neptune	Solar System
Small Clouds	Large Clouds	Moon of Saturn: Titan	Planet Uranus	Supernovas
Satellites	Pluto	Planet Mars	Planet Saturn	Nebulas
Space Shuttle	Moon of Mars: Phobos	Planet Venus	Planet Jupiter	Galaxy
International Space Station				Universe
Most Meteors	Some Meteors		Some Stars	Some Stars
Some Asteroids	Some Asteroids			
Probably Some Black Holes	Some Black Holes	Probably Some Black Holes	Probably Some Black Holes	Some Black Holes
Some Com-ets	Most Com-ets	Some Comet Tails	Some Comet Tails	Some Comet Tails

OPTIONAL PROMPTS FOR WRITING OR DISCUSSION

You may want to choose one or more of the prompts below for science journal writing during class or as homework. Or the prompts could be used for a discussion or during a final student sharing circle in which each student gets a turn to share.

- Tell how successful you and your partner were with your sort and why.
- Tell one thing you learned about the size of sky objects that you didn't know before today.
- Make a list of 20 space objects, from smallest to largest.

The Sizes of Sky Objects Chart

1. Add to key concept chart. When you have finished the tour, have students note that even though the Sun is super huge, many objects in space are even bigger. Post the following key concept on the concept wall:

> There are many things in the Universe that are much larger than the Sun.

2. Project the *Sizes of Sky Objects* chart. After the tour, show the *Sizes of Sky Objects* transparency for students to use to double-check their revised card sorts. Ask if any surprised them.

3. Which objects are not in space? Tell them that almost all these objects are in space. Ask if they can name the objects that are not in space, but are in Earth's atmosphere. [Birds, meteors, and clouds.]

Unit Goals

Size: Some sky objects are relatively small, and some are huge.

Distance: Some objects are relatively close to Earth and some are very far away.

Distance of sky objects from us affects their apparent size: Large objects appear small when far away.

TEACHER CONSIDERATIONS

PROVIDING MORE EXPERIENCE

Questions about planet sizes: Have students consult the *Visual Chart of Sizes of Planets in the Solar System* sheet. Ask the following questions to help them focus on the relative sizes of planets in our Solar System:

- What planet is almost the same size as Earth? [Venus.]
- What other planets are smaller than Earth? [Mercury and Mars.]
- What planets are bigger than Earth? [Jupiter, Saturn, Uranus, and Neptune.]
- What is the biggest object in the Solar System? [Sun.]

Mnemonic Devices to Remember Planet Names: Mnemonic devices are tricks used to help remember something. If you'd like your students to memorize the names and order of the planets in the Solar System, you may choose to share with them the following mnemonic device:

My **V**ery **E**ducated **M**other **J**ust **S**erved **U**s **N**oodles.

Mercury, Venus, Earth, Mars, Jupiter, Saturn, Uranus, Neptune. Alternatively, you could have students generate their own original mnemonic devices.

Categorizing Their Own Drawings by Actual Size: Post large signs on the wall with the same categories used in the activity, plus *Not Sure.* Leave the transparency of the Sizes of Sky Objects chart up on the overhead projector as a reference.

Tell students that they will categorize their drawings from Session 1.2 according to the real sizes of the objects. They will be responsible for putting their own drawings wherever they think they belong. Point out the extra category for *Not Sure.*

To avoid crowding, assign a few students at a time to get their drawings and put them on the wall in the appropriate categories.

When all students have finished, hold a discussion. Ask for help in placing objects that ended up in the *Not Sure* category. For objects that no one is sure about, suggest further research. Ask if anyone has any concerns about other objects placed in categories, and if so, have them explain why. Again, suggest further research if necessary.

Key Vocabulary

Science and Inquiry Vocabulary

Evidence

Scientific Explanation

Model

Scale Model

Prediction

Scientist

Three–Dimensional (3-D)

Two–Dimensional (2-D)

Space Science Vocabulary

Atmosphere

Satellite

Orbit

Diameter

Sphere

System

SESSION 1.7

How Far Away Are They?

**Note: The 13 station signs from this session will be combined in Session 8 to make a single tall poster that compares the distances with all the sky objects, from the ground to the International Space Station.*

Unit Goals

Size: Some sky objects are relatively small, and some are huge.

Distance: Some objects are relatively close to Earth and some are very far away.

Distance of sky objects from us affects their apparent size: Large objects appear small when far away.

Overview

In previous sessions, the students compared the sizes of various sky objects and began to explore how distance affects their apparent size. In Sessions 1.7 and 1.8, students use models to measure how far away these sky objects are. Session 1.7 focuses on distances of objects between the Earth and the Moon; in Session 1.8, students contrast them with the relatively vast distance to the Sun.

Using a *new* scale ruler, student pairs circulate to learning stations around the classroom to measure the distance between the ground and 13 different objects—including the tallest mountain, a cloud, a hot air balloon, an airplane, satellites, the International Space Station, and the Moon. They make their measurements on scale drawings.

The class discusses their findings. The relative distances of all objects are discussed, including how much farther away the Moon is than any of the other objects measured. They also discuss which objects are in Earth's atmosphere and which are in space.

How Far Away Are They?	Estimated Time
Introducing measuring distances	15 minutes
Going to the learning stations	35 minutes
Discussing distances	10 minutes
TOTAL	**60 minutes**

What You Need

For the class

- ❑ 1 copy of each of 13 station signs from the student sheet packet: Tallest Tree, Tallest Building, High-Flying Bird, First Hot Air Balloon with Passengers, Cloud, Airplane, Highest Skydiver, Tallest Mountain, Satellite #2, Moon, Top of Atmosphere, Satellite #1, International Space Station
- ❑ 1 copy of the 3 "ground" sheets that go with 3 of the station signs (Top of Atmosphere, Satellite #1, International Space Station), from the student sheet packet. (See Getting Ready, #3 which follows.)
- ❑ 1 roll masking tape
- ❑ sentence strips for 2 key concepts
- ❑ wide-tip felt pen

For each student

- ❑ *How High in the Sky?* from the student sheet packet
- ❑ pencil

TEACHER CONSIDERATIONS

TEACHING NOTES

Changes of scale: Because of the vast sizes and distances in space compared with our surroundings on Earth, any comparison between them requires changes in scale. We have limited the number of times we change the scale of the models that students use during the unit. In Session 1.4, students used a scale ruler where 1 mm=3,000 km in order to measure the sizes of the Earth, Moon, and Sun. That same scale will be used in Session 1.8 to measure the distances among these objects. However, students need a different scale ruler in Session 1.7 to compare the distances among such sky objects as birds, airplanes, and satellites. For this session, students use a scale ruler on which 1 centimeter represents 2 kilometers.

What Some Teachers Said

"The students enjoyed visiting the various stations to measure the distance of different space objects from Earth. They made great discoveries and were overwhelmed by the numbers."

"At one point during this lesson, I watched my students as they busily measured each one of the stations. They were completely engaged. (Wish I would have had my camera.)"

"If a stranger had walked into the room, they would have seen all of the students busily involved on the tasks and moving around from station to station. They were very interested in the activity."

Getting Ready

Preparing the station signs
In this session, pairs of students go to learning stations to measure the distance from the ground to 13 sky objects. Each object is pictured on a station sign, and students do the measuring on the signs. You will need to provide eight blank 8½ X 11 inch pieces of paper to serve as extensions for three of the signs. (See #3, which follows.)

1. Photocopy the station signs. Photocopy 13 station signs and the three "ground" sheets that go with 3 of them (Top of Atmosphere, Satellite #1, and International Space Station).

2. Post the 10 single-page signs. Tape the 10 signs in the list that follows to the wall around the classroom in places that will be accessible to students, spaced so that crowding and traffic problems will be minimized. They don't need to be in any particular order.

1. *Cloud*
2. *Highest Skydiver*
3. *First Hot Air Balloon with Passengers*
4. *Airplane*
5. *Tallest Building*
6. *Tallest Tree*
7. *Tallest Mountain*
8. *Bird*
9. *Moon*
10. *Satellite*

3. Post the three multipage signs. Three of the signs need to be constructed of more than one page, taped together vertically in a column, as outlined in the following. Choose wall space for these three signs where you can start the column near the floor, so students can measure from the "ground" on the sign to the sky object pictured.

- *Top of atmosphere.* This station sign will be made up of three sheets. On the bottom is a "ground" sheet, followed by a blank sheet, followed by the Top of Atmosphere sheet.
- *Satellite #1.* This sign has a total of five sheets taped together vertically. From bottom to top, they are "ground" sheet, followed by three blank sheets, followed by the Satellite #1 sheet.
- *International Space Station.* This sign has a total of seven sheets: the ground, followed by five blank sheets, followed by the International Space Station sheet.

Unit Goals

Size: Some sky objects are relatively small, and some are huge.

Distance: Some objects are relatively close to Earth and some are very far away.

Distance of sky objects from us affects their apparent size: Large objects appear small when far away.

TEACHER CONSIDERATIONS

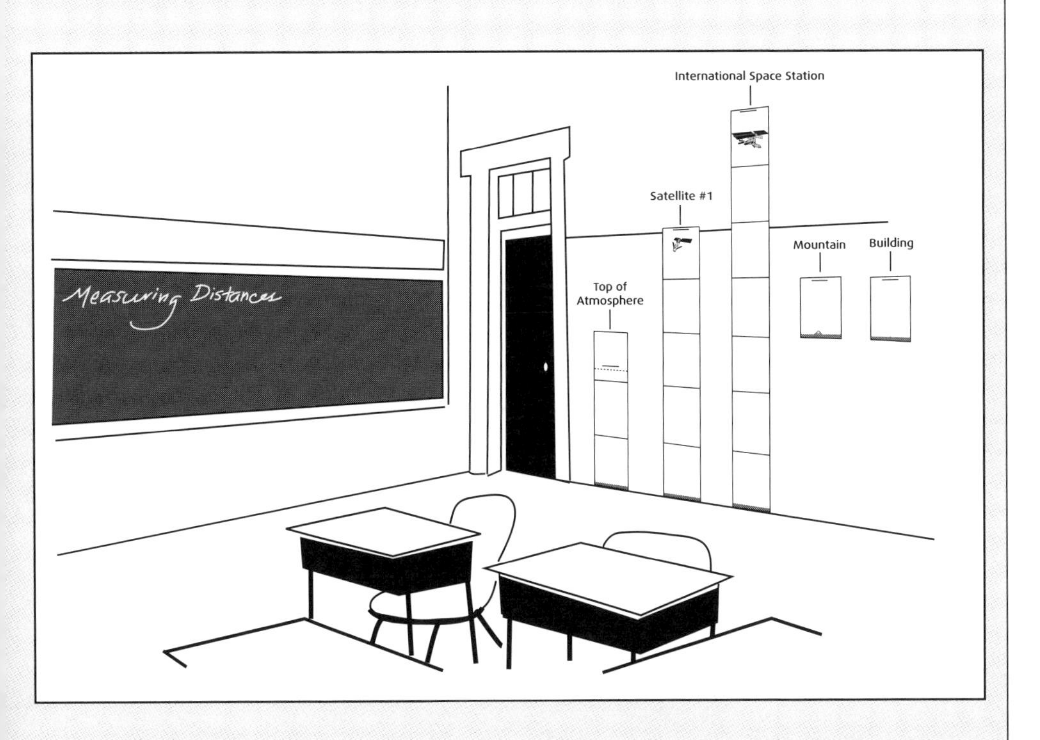

4. *Optional: Make extra signs.* If you have a large class, you may want to make duplicates of one or two of the station signs. This way, there will be less crowding as students measure. We suggest making duplicates of signs for Satellite #1 and the International Space Station.

5. Copy the student sheet. For each student, make one copy of the *How High in the Sky?* student sheet.

6. Locate landmarks. Pace out 150 meters from your classroom, and find an easily recognizable landmark that you can describe for your students to represent the distance to Satellite #2. Also, refresh your memory about the landmarks you chose for earlier sessions that are about 1 km and 2 km from your school.

7. Write the following two key concepts on sentence strips and have them ready to post on the concept wall:

Some objects we see in the sky, such as birds and airplanes, are relatively close.
The Moon is very far from us compared with objects in Earth's atmosphere.

Introducing Measuring Distances

1. Evidence about distances to sky objects. Ask students how far away they think clouds are. The top of the tallest mountain? The beginning of space? Remind them that great thinkers change their ideas when they find evidence that doesn't match their ideas. Ask what kinds of evidence might make them change their minds about how far away these things are. [They will probably bring up measurement.]

2. We've measured sizes; now distances. Remind the class that they have been learning about the sizes of sky objects. Tell them that in this session, they will learn how far away they are. For example, they will find out what the distance is from the ground to a cloud, an airplane, a satellite, or the Moon. Tell them that in the next class session, they'll measure distances beyond the Moon.

3. Review meter and kilometer. Ask students to show with their hands about how long a meter is. Ask how long a kilometer is. Remind them of the local landmark 1 kilometer away that you told them about in an earlier session. Tell them that they'll be measuring in meters and kilometers how high sky objects are, straight up from the ground.

Unit Goals

Size: Some sky objects are relatively small, and some are huge.

Distance: Some objects are relatively close to Earth and some are very far away.

Distance of sky objects from us affects their apparent size: Large objects appear small when far away.

TEACHER CONSIDERATIONS

TEACHING NOTES

Introduce Altitude: You may want to use the word *altitude* during this session. Altitude can be defined as the height, or elevation, above sea level on the Earth's surface.

Note: For your own background information, and not for students of this age range, in a more technical astronomical sense, *altitude* can also refer to the angle between the horizon and an object in the sky.

.

Key Vocabulary

Science and Inquiry Vocabulary

Evidence

Scientific Explanation

Model

Scale Model

Prediction

Scientist

Three–Dimensional (3-D)

Two–Dimensional (2-D)

Space Science Vocabulary

Atmosphere

Satellite

Orbit

Diameter

Sphere

System

4. Measuring distance models. Let them know that because it isn't practical for them to measure the real distances, they'll be measuring models. The models are drawings based on measurements that scientists have made. Review the related key concepts on the concept wall about how scientists use models.

5. New scale rulers: 1 cm = 2 km. To measure these distances, they will get another special scale ruler. Tell them that, on this ruler, 1 centimeter will represent 2 kilometers.

Going to the Learning Stations

1. Student data sheet. Pass out a *How High in the Air?* student sheet to each student. Tell students that they will use the scale ruler on the edge of the sheet to measure. Mention that they will not cut off the ruler: they will just hold the edge of the sheet next to each object that they measure.

2. Introduce the measurement learning stations. Point out the stations around the room, each with a scale model drawing of how high an object is. Tell them that they will go to each station in pairs to measure the distance from the ground to the object, and record it on their student sheet in meters or kilometers.

3. Measure from the "ground" line to the object. Go to a station they can see from their seats, such as Tallest Mountain, and demonstrate briefly how to measure from the ground line to the top of the object pictured. At other stations, measure to either the top or the bottom of the object, according to what the station directions say.

4. What is the "top" of the atmosphere? Review, "What is Earth's atmosphere?" [In Session 1.1, students learned that the Earth is surrounded by a layer of gases, or "air," which is called the atmosphere.] The higher you go from the ground, the less air there is, until there's no air at all. Tell them that there is no exact line where the atmosphere ends and space begins—instead, it just slowly fades out. For this reason, it is hard to say exactly how high the atmosphere is. Tell them that when they measure the top of the atmosphere, they will be measuring to a point in the sky where there is very little air.

5. Some distances are too high to measure, even in this scale model. Tell them that some of the objects are so high that they will not be able to measure them, even in this scale model. Tell them that for these objects they will find the distance written on the sheet, and they will just write it down on their student sheets.

Unit Goals

Size: Some sky objects are relatively small, and some are huge.

Distance: Some objects are relatively close to Earth and some are very far away.

Distance of sky objects from us affects their apparent size: Large objects appear small when far away.

HOW HIGH IN THE AIR?

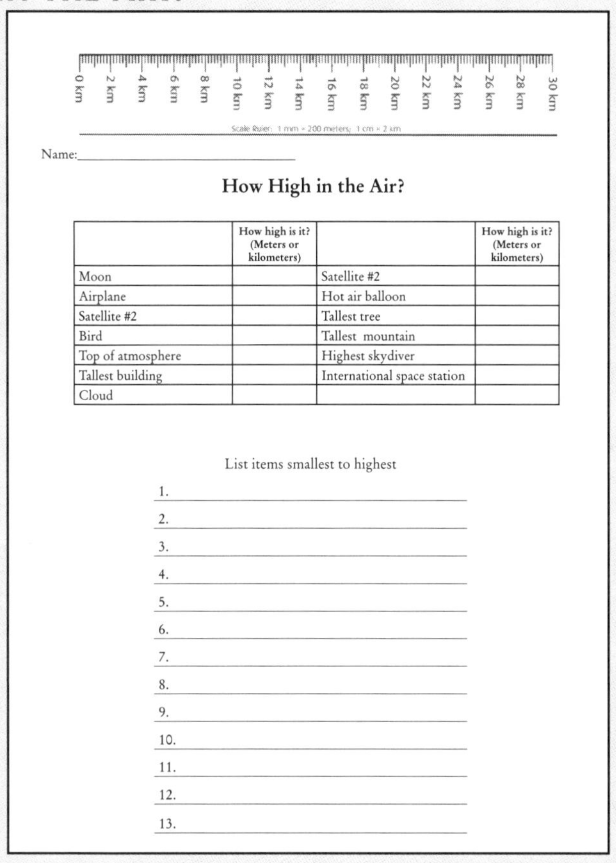

0 km | 2 km | 4 km | 6 km | 8 km | 10 km | 12 km | 14 km | 16 km | 18 km | 20 km | 22 km | 24 km | 26 km | 28 km | 30 km

Scale Ruler: 1 mm = 200 meters; 1 cm = 2 km

Name:______________________________

How High in the Air?

	How high is it? (Meters or kilometers)		How high is it? (Meters or kilometers)
Moon		Satellite #2	
Airplane		Hot air balloon	
Satellite #2		Tallest tree	
Bird		Tallest mountain	
Top of atmosphere		Highest skydiver	
Tallest building		International space station	
Cloud			

List items smallest to highest

1. ______________________
2. ______________________
3. ______________________
4. ______________________
5. ______________________
6. ______________________
7. ______________________
8. ______________________
9. ______________________
10. ______________________
11. ______________________
12. ______________________
13. ______________________

6. Explain duplicate stations. If you made a duplicate of any of the stations (suggested for Satellite #1 and the International Space Station), tell the students that you did this so there would be enough stations to keep them from getting crowded. Tell them which stations are identical, and not to go to both.

7. Show how to record on the student sheet. Tell them that there are 13 sky objects listed on their student sheets, and they can go to the stations in any order. If a station is crowded, they should go there later. Explain your expectations for cooperative behavior as they move around the stations.

**Note: See the chart on page 197 for correct measurements and actual distances of objects.*

8. Ranking the distances on their student sheets. Say that when they have finished, they will return to their seats and fill out the bottom half of their sheets. If necessary, go over how they will use their measurements to put the objects in order, from closest to the ground to farthest away. Assign teams of two and have them go to the stations and begin.

Discussing Distances

1. Questions about distances of the objects. When students have returned to their seats and finished ranking objects on the bottom half of their sheets, get the attention of the class. Ask the following questions:

a. Which object was the greatest distance from the ground? [Moon.] The shortest distance? [Top of tallest tree.]

b. Did you find anything surprising about the distances you measured? Point out that although the tallest tree might look very high to us, it is not very high at all when compared with other things that are much higher.

c. If someone was on top of Mt. Everest, would airplanes be flying above or below them? [They could be flying above *or* below them.]

d. Do airplanes ever fly above clouds, or are clouds too high? [Airplanes can fly higher than clouds.]

e. Are satellites above or below clouds? [*Much* higher than clouds. Satellite images of the Earth show clouds far below.]

f. Do birds ever fly as high as Mt. Everest? [Yes.] Higher than clouds? [Yes.]

Unit Goals

Size: Some sky objects are relatively small, and some are huge.

Distance: Some objects are relatively close to Earth and some are very far away.

Distance of sky objects from us affects their apparent size: Large objects appear small when far away.

TEACHER CONSIDERATIONS

APPROXIMATE DISTANCES USED IN THE MODEL DRAWINGS AT LEARNING STATIONS

OBJECT	DISTANCE IN MODEL	ACTUAL DISTANCE
Tallest Tree	1 mm	112 m
Tallest Building	2 mm	450 m
Sheep/Duck/Chicken Balloon	1 cm	2 km
Cloud	3 cm	6 km
Tallest Mountain	3.5 cm	7 km
Bird	5 cm	10 km
Airplane	5 cm	10 km
Highest Skydiver	15.5 cm	31 km
Top of Atmosphere	60 cm	120 km
Satellite #1	125 cm	250 km
International Space Station	175 cm	350 km
Satellite #2	160 m (bigger than schoolyard)	32,000 km
Moon	2 km (more than 12 blocks away)	385,000 km
Sun	750 km (compare with width of whole state)	150,000,000 km

Key Vocabulary

Science and Inquiry Vocabulary

Evidence

Scientific Explanation

Model

Scale Model

Prediction

Scientist

Three–Dimensional (3-D)

Two–Dimensional (2-D)

Space Science Vocabulary

Atmosphere

Satellite

Orbit

Diameter

Sphere

System

2. The Moon is much farther than the other objects measured. Point out how much farther away the Moon is than any of these other objects (about 385,000 km, which, in this model, would be about 2 km—more than 12 blocks—away). The Moon is very far away!

3. Satellite questions. Ask the following questions:

- Objects in space or the atmosphere? Ask what their measurements were at the Top of the Atmosphere station. [About 120 km.] Ask, "Which objects are above the Earth's atmosphere?" [According to their measurements, Satellites #1 and #2, the International Space Station, and the Moon are all beyond the atmosphere.] Attach a sign, *Space,* somewhere above where they measured the top of the atmosphere.

- Different definitions of the end of the atmosphere (120 km or 500 km). Tell students that because the atmosphere gets thinner gradually, some scientists might say the atmosphere ends much farther away than 120 km, where they measured. Point to a place about 2 1/2 meters high on the wall. Say at this scale, this is about 500 km above the ground, where some scientists say the Earth's thinnest, outer atmosphere ends and space begins. According to the 500–km definition, Satellite #1 and the International Space Station are *within* the thin part of the upper atmosphere, but there is so little air there, it's pretty nearly no atmosphere at all.

- Which objects are in Earth's atmosphere? [All the rest of the objects they measured.]

4. Satellites orbit Earth very high. Point out that Satellite #1 is 25 times higher than an airplane! Other satellites, such as Satellite #2, are much higher than that.

5. Seeing satellites in the night sky. Tell them that sometimes they can see satellites in the night sky. A satellite looks like a moving white dot as it orbits the Earth. Ask if any students think they have seen a satellite. Emphasize how amazing it is to be able to see something so small and so much higher than an airplane without a telescope.

Unit Goals

Size: Some sky objects are relatively small, and some are huge.

Distance: Some objects are relatively close to Earth and some are very far away.

Distance of sky objects from us affects their apparent size: Large objects appear small when far away.

TEACHER CONSIDERATIONS

SCIENCE NOTES

Some additional interesting facts: Students don't need to know this, but in case they ask, at this scale, the Earth would be 65 m in diameter, the Moon would be 15.25 m in diameter, and the Sun would be 7 km in diameter. The Sun in this model would be 750 km away.

At 1,000 km per hour (the speed of a fast passenger jet), it would take sixteen days to get to the Moon. To travel the distance that the Sun is from us, at 1,000 km/hour, would take more than 16 years!

The "Top" of Earth's Atmosphere: Because the edge of the atmosphere is a continuum of thinner and thinner gases, there is no such thing as a single point where the atmosphere ends and space begins. This is why different resources list different heights for the end of the atmosphere. They are all somewhat arbitrary points. For this activity, we have chosen to mark the top of the atmosphere at 120 km. During the class discussion after this activity, students will find that, depending on how you define the edge of the atmosphere, it could be said to extend up to 500 km from the Earth (2.5 meters in this model) or more, well past the International Space Station.

Why do satellites keep orbiting the Earth? Why don't they drift away into space? You may want to allow an opportunity for students to discuss what they might know or wonder about gravity. Tell them they'll think and learn more about gravity in Unit 2.

Key Vocabulary

Science and Inquiry Vocabulary

Evidence

Scientific Explanation

Model

Scale Model

Prediction

Scientist

Three–Dimensional (3-D)

Two–Dimensional (2-D)

Space Science Vocabulary

Atmosphere

Satellite

Orbit

Diameter

Sphere

System

6. Key Concepts. Point out that they now have a good sense of how far from Earth various sky objects really are. Post the two key concepts for this session.

Some objects we see in the sky, such as birds and airplanes, are relatively close.
The Moon is very far from us compared with objects in the Earth's atmosphere.

Unit Goals

Size: Some sky objects are relatively small, and some are huge.

Distance: Some objects are relatively close to Earth and some are very far away.

Distance of sky objects from us affects their apparent size: Large objects appear small when far away.

TEACHER CONSIDERATIONS

OPTIONAL PROMPT FOR WRITING OR DISCUSSION

You may want to have students use the prompt below for science journal writing during class or as homework. Or it could be used for a discussion or during a final student sharing circle.

- Name one or more things you learned today about how high objects are in the sky.

Key Vocabulary

Science and Inquiry Vocabulary

Evidence

Scientific Explanation

Model

Scale Model

Prediction

Scientist

Three–Dimensional (3-D)

Two–Dimensional (2-D)

Space Science Vocabulary

Atmosphere

Satellite

Orbit

Diameter

Sphere

System

SESSION 1.8

Comparing Distances

Overview

Students begin by reading *Jumping from the Edge of Space,* which is the story of the first person to ride a helium balloon to "the edge of space" and skydive back to Earth. The class puts the altitude of the skydive in the context of a wall chart combining the relative distances of the sky objects they measured in Session 1.7. The reading gives students another opportunity to think about relative distances, to visualize Earth's atmosphere, and to understand the idea that air gets thinner the higher you go, until there is no longer any air. It is also another example of how new exploration has provided evidence to answer questions about space science.

Using the scale from Session 1.4, the class compares the distance to the Moon with the other distances they've measured. The relatively small distance to the top of Earth's atmosphere is evident when contrasted with the distance to the Moon at this scale. Students go outside (or into a long hallway or large room) to pace off the distance to the Sun, and contrast this distance with the distance to the Moon.

Back in the classroom, students discuss a question about the relative distances of the Moon and Sun in small groups called *evidence circles.* They apply their new evidence as they discuss the question.

Comparing Distances	Estimated Time
Reading and discussing *Jumping from the Edge of Space*	15 minutes
The distance between the Earth and Moon	10 minutes
Measuring the distance to the Sun	10 minutes
Evidence circles: discuss relative distances of Earth, Moon, and Sun	15 minutes
Large-group discussion of where to put Moon in the model	10 minutes
TOTAL	**60 minutes**

What You Need

For the class

- ❑ station signs from previous session
- ❑ 3-D models of Sun, Earth, and Moon from Session 1.4
- ❑ 1 copy of *Multiple Sky Objects* station sign, from the student sheet packet
- ❑ wide-tip felt pen
- ❑ sentence strips for 2 key concepts
- ❑ *optional:* computer and large screen monitor or LCD projector on which to show CD-ROM file of Kittinger's Jump to the class

Unit Goals

Size: Some sky objects are relatively small, and some are huge.

Distance: Some objects are relatively close to Earth and some are very far away.

Distance of sky objects from us affects their apparent size: Large objects appear small when far away.

TEACHER CONSIDERATIONS

CD-ROM NOTES

Optional Skydive Interactive: Jump from the edge of space in a game that follows the story of Captain Joseph Kittinger's record-breaking jump of over 100,000 feet. Students are introduced to Captain Kittinger's mission and navigate through the story using the "back" and "next" buttons. At the end of the introduction, students can continue to the game by clicking "Try your own jump."

The game will start by simulating Kittinger's climb through the atmosphere to 100,000 feet. The panel at the right displays the altitude to which Kittinger's gondola has climbed. Students should pay attention to the story and to the atmosphere descriptions on the left panel, as information in the panel will assist them in deploying the chutes correctly. Once the climb is complete, students should press the jump button at the bottom left of the screen to initialize the jump sequence.

During the jump, students will need to deploy their stabilizing chute between 90,000 and 80,000 feet by pressing the "Deploy Stabilizing Chute" button at the bottom center of the screen. Students should deploy their main chute below 20,000 feet by pressing the "Deploy Main Chute" button. The buttons on the right side of the screen allow the student to pause the climb or jump by pressing the "pause." By pressing the "Start Over" button or the "Quit" button students can return to the introduction of Kittinger's mission. After the jump is completed a screen will appear allowing the student to play again.

Further instructions for using this program are included on the CD-ROM.

Key Vocabulary

Science and Inquiry Vocabulary

Evidence

Scientific Explanation

Model

Scale Model

Prediction

Scientist

Three–Dimensional (3-D)

Two–Dimensional (2-D)

Space Science Vocabulary

Atmosphere

Satellite

Orbit

Diameter

Sphere

System

For each student
- ❑ 1 copy of the sheet *Evidence About Distances Among the Earth, Moon, and Sun,* from the student sheet packet
- ❑ 1 copy of the reading, *Jumping from the Edge of Space,* from the student sheet packet

**Note that the second page of the reading is optional, but strongly recommended. It can be photocopied on the back of the first page. Additional pages of the reading are on the CD-ROM.*

Getting Ready

1. *Optional but strongly recommended:* If possible, prepare to show the CD-ROM of Kittinger's Jump to the whole class or plan for students to rotate to view it on a computer.

2. Make copies of the reading *Jumping from the Edge of Space* for each student. *Optional: Copy page 2 of the reading on the back of the first page.*

3. Make copies of the sheet *Evidence About Distances Among the Earth, Moon, and Sun* for each student.

4. Make one copy of the *Multiple Sky Objects* station sign.

5. Combine the station signs from Session 1.7 into one long column on the wall. To do this, you will layer three station signs together as follows. Leave the I*nternational Space Station* sign on the wall. Tape the *Satellite #1* sign on top of it as illustrated. Next, tape the *Top of the Atmosphere* on top of that. Finally, add the *Multiple Sky Objects* sign as shown on page 205.

6. Decide where you will take the class to pace off the 48–meter distance to the Sun.

7. Write the following key concepts on sentence strips, and have them ready to post on the concept wall:

The atmosphere is thin compared with the size of the Earth—like the skin of an apple.
The Sun is much farther from Earth than the Moon is.

Unit Goals

Size: Some sky objects are relatively small, and some are huge.

Distance: Some objects are relatively close to Earth and some are very far away.

Distance of sky objects from us affects their apparent size: Large objects appear small when far away.

TEACHER CONSIDERATIONS

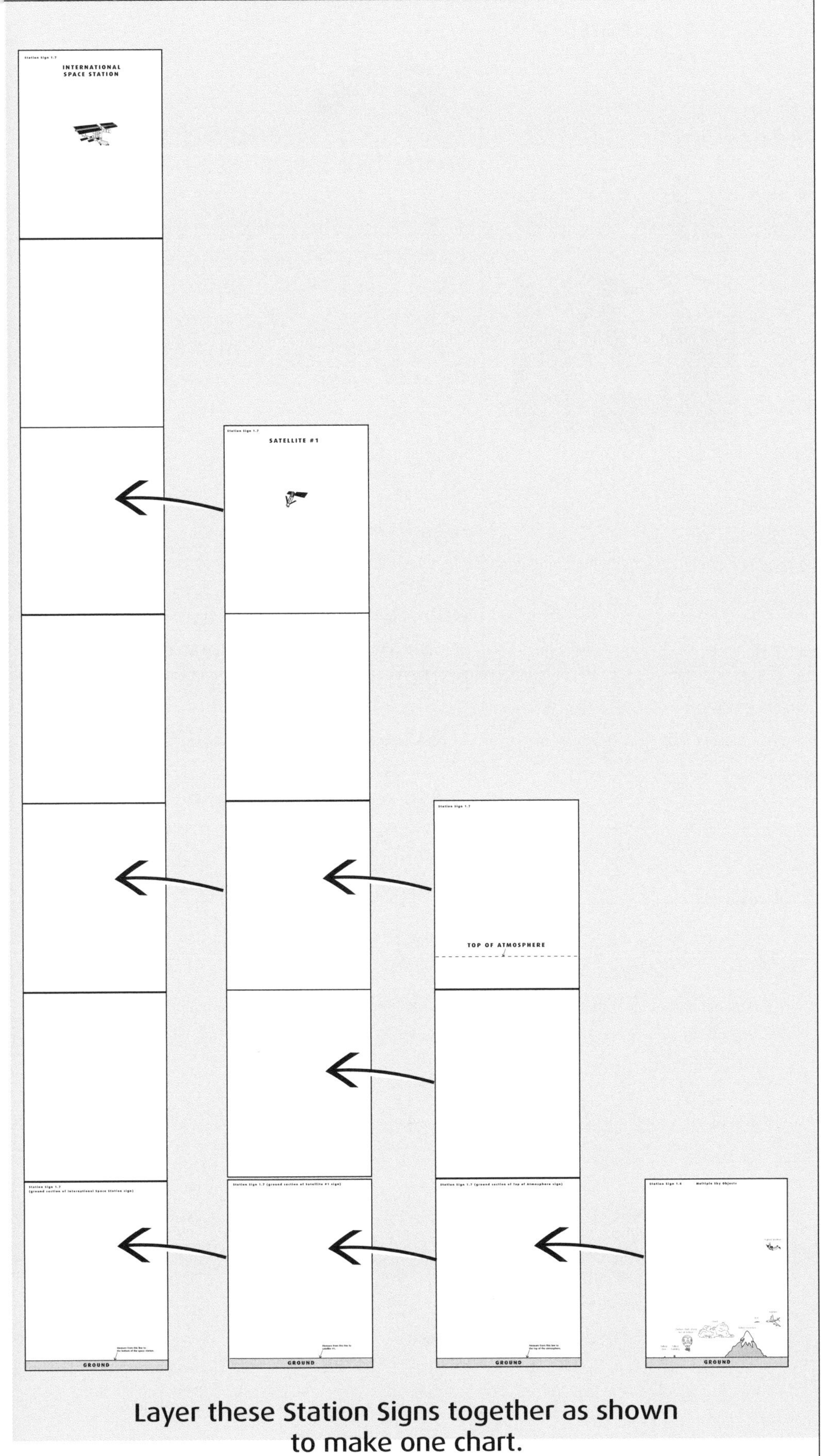

Layer these Station Signs together as shown to make one chart.

INTERNATIONAL SPACE STATION

SATELLITE #1

TOP OF ATMOSPHERE

GROUND

Resulting combined chart.

JUMPING FROM THE EDGE OF SPACE, PAGE 1

Name:________________________

Jumping from the Edge of Space

In 1960, a man jumped from the edge of space back to Earth. His name was Joseph Kittinger. Back then, scientists wanted to find out if people could work in space, where there is no air. They also wanted evidence that people could get back to Earth with a parachute if something bad happened.

Joseph went up in a balloon filled with light gas—the same kind of gas that is in birthday balloons. The balloon was 60 meters in diameter. The higher you travel in the atmosphere, the less air there is. Fire needs air to burn, so hot air balloons, which need fire, cannot travel too high. His balloon took him about 31,000 meters (31 km) high. He was wearing a space suit, which gave him air to breathe.

Photographs courtesy of the National Museum of the United States Air Force

At about seven in the morning, Joseph stepped out and fell toward Earth. When people parachute from lower in the atmosphere, the air slows them down and makes a lot of noise as they fall through it. But because there was very little air that high in the sky, Joseph fell very quickly and very quietly.

Far below, Joseph could see clouds. Because he was above the clouds and it was daytime, the Sun was shining on him. As he was falling, he rolled over and looked up. Although it was morning, the sky looked black.

Joseph kept falling and falling and falling. He fell at 990 km per hour (615 miles per hour), close to the speed of sound. After four minutes of falling, his main parachute opened and slowed him down. He landed on Earth in the middle of the desert 13 minutes and 45 seconds after jumping out of the balloon.

Many years have passed since Joseph made his jump, but no one else has ever jumped from that high and survived. Another man tried, but he died. There are other people who hope to break Joseph's record. Maybe by the time you read this, one of them will have done it.

Was Joseph the first person in space? No, but he was the first to jump from the edge of space. Still, there is no line where the atmosphere ends and space begins. There is just less and less air, until there is no air. A balloon full of light gas can rise even in very little air, but it cannot keep going up if there is no air. To go beyond the air in the atmosphere, you need a space ship with rockets. Other people did that later, but Joseph traveled to the edge of space and jumped.

GO!

Reading: Jumping from the Edge of Space

1. Reading about highest skydive jump. Tell students they're going to read about a person who jumped from the "edge of space." Depending on students' reading abilities, you may want to have them do independent, paired, or shared reading.

Note: You may want to review some or all of these words before the reading: *parachute, breathe, rocket.*

2. *Optional*: Early finishers read second page. Say that if they finish reading on the first page, they should continue reading the back of the page.

Discussing the Reading

1. Ask students for comments. After the reading, allow a few minutes for students to comment. There will likely be many comments and questions, but if not, you might prompt students by asking:

- What was different high in the sky than on the ground? [Colder, much less air, much less air pressure, black sky above, no sound when falling, and so on.]
- How do you think things looked as he fell and got closer and closer to Earth?

2. Put Kittinger's jump in the context of other objects measured. When students bring up the height (or altitude) of the jump, draw their attention to the composite chart that you made from the station signs, and point out that it is really a scale model showing the distances to sky objects. As they look at the chart, ask:

TEACHER CONSIDERATIONS

TEACHING NOTES

Time allotted for discussion: In order to keep this session a reasonable length, only 15 minutes have been allotted for the *Jumping From the Edge of Space* story and discussion. Please note that your students will likely be excited about discussing other aspects of this amazing story. If you have available time, you may choose to allow more discussion. You may also choose to have your students read a fascinating interview with Joseph Kittinger and additional information about him found on the CD-ROM.

Reading level: The reading level of page 1 of the reading is appropriate for most third and fourth graders. Page 2 and the additional pages on the CD-ROM are for students who are interested in further information on the topic, and who are able to read at a slightly higher level.

Purposes of the Reading:

- Adds to the idea of space exploration: people sending spacecraft and themselves up into the sky to explore, and seeking evidence. This follows *The Adventures of a Sheep, a Duck, and a Chicken*, in the continuing story about different levels of space exploration.
- Helps students visualize the layer of air (or atmosphere) around our planet.
- Reinforces the understanding that there is air in our atmosphere, but that the air thins as you go higher toward space. Helps prepare students to understand that there is no air in space.
- Many students think that gravity is caused by or somehow dependent on air. When they later learn that there is gravity on the Moon and on other planets, even when there is no air, it helps to address this misconception.
- Metric measurement application with the diameter of the balloon and the height of the flight.
- Page 2 can be used to discuss how misinterpretation of observations can result in mistaken ideas—such as aliens landing.

JUMPING FROM THE EDGE OF SPACE, PAGE 2

Name:______________________________

Jumping from the Edge of Space

- Joseph Kittinger set records for the highest ride in a balloon and for jumping from the greatest height. Setting records was not the reason for his jump, though. He was trying to find evidence about what would happen to a person high in the atmosphere and whether people could safely jump back to Earth from that height.
- Others are trying to set new records by riding and jumping from even higher in the atmosphere. One of the teams will use a balloon as tall as the Empire State Building. The skin of the balloon is so thin that it will be hard to launch it without ripping the skin.
- In 1999, Bertrand Piccard and Brian Jones made the first nonstop balloon flight around the world. It took them 19 days, 21 hours, and 55 minutes.

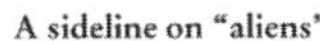

A sideline on "aliens"

- Joseph's jump was very dangerous, and scientists were not sure if he could do it and live. Before he tried it, people took plastic dummies with parachutes up in a balloon and pushed them out. This was near Roswell, New Mexico. Some people say they saw aliens landing in that area around the same time. Experts say that people who saw the dummies fall thought that they were aliens from another planet.
- One of the men Joseph was working with got hit in the head by the gondola that hangs under a balloon. The man's head became very swollen. Joseph helped his friend walk into the hospital. Some people who thought they saw aliens in Roswell said they saw a red-haired man walking an alien with a huge head into the hospital. Joseph has red hair. He thinks the people saw him and his friend with the swollen head.

Photographs courtesy of the National Museum of the United States Air Force

a. How high did Kittinger go before he jumped? [About 31 kilometers (31,300 m).] Tell them that this would be about 15 1/2 cm in the chart's scale model. Point out the drawing of Kittinger's jump on the chart.

b. Which objects that we measured are below this level? [Tallest tree; tallest building; the Sheep, Duck, and Chicken balloon; cloud; tallest mountain; bird; and airplane.]

c. Which objects that we measured are above this level? [The top of the atmosphere, Satellite #1, Space Station, Satellite #2, and the Moon.]

d. Did Kittinger really jump from the "edge of space?" [In Session 1.6, students measured 120 km to the "top of the atmosphere." In Session 1.7, they learned that some scientists say that space begins at about 500 km. By either definition, Kittinger was within the atmosphere when he jumped.] Point out that even though Kittinger jumped from very high up, there isn't really an "edge" of space, so it's hard to say exactly where space begins, but 31 km is very high in the atmosphere.

e. Could Kittinger breathe up there? [Not without air tanks.] Review that we need air to survive, and there is less air as you go higher. Beyond the atmosphere, in space, there is no air. Mention that the Moon has no atmosphere, but some planets and moons do.

Changing the Scale of the Model To Include the Moon and Sun

1. How far away is the Sun? Ask, "In relation to this scale model chart, where do you think the Sun is?" Accept all answers. Tell students that they are going to measure the distance to the Sun compared with these other distances that they have measured.

2. Need a different scale model. Remind students that in Session 1.7, they couldn't measure the distance to Satellite #2 or the Moon because they wouldn't fit in the classroom. That scale model was not practical for students to measure these longer distances, so today they will change that scale model to better understand and compare the distances. Tell them that scientists change models to fit their needs.

3. The 3-D Sun, Earth, Moon models. Hold up the 3-D Sun, Earth, and Moon models used in Session 1.4, and say that they're going to use this scale. Remind them that these are scale models of the sizes of the three objects compared with one another.

Unit Goals

Size: Some sky objects are relatively small, and some are huge.

Distance: Some objects are relatively close to Earth and some are very far away.

Distance of sky objects from us affects their apparent size: Large objects appear small when far away.

TEACHER CONSIDERATIONS

OPTIONAL PROMPTS

Additional suggested discussion questions for after the reading:

- Why did Joseph Kittinger make the jump? [Joseph Kittinger and his team were seeking evidence to answer the questions that follow.]

  —Is it possible to skydive safely from a spacecraft that might be in trouble near the edge of Earth's atmosphere?

  —What are the effects of high altitude on people?

- What kind of gas was in the balloon? [Helium.]

- Why didn't they use a hot air balloon? [Hot air balloons can't fly that high because the air is so thin that the balloon would need to be very hot to stay bouyant.]

- Why do you think the sky looked black when he looked up? [For more information see optional additional pages of the student reading on the CD-ROM.]

- Why can't balloons travel all the way into space?

- What problems do you think people who are trying to break his record might have to deal with? [With a balloon as tall as the Empire State Building, and with very thin fabric, wind and tearing fabric are a big concern.]

- How do you think it would feel to jump from so high?

TEACHING NOTES

In this 3-D model: The Sun is 44 cm in diameter. The Earth and Moon are 4 mm and 1 mm in diameter, respectively.

Key Vocabulary

Science and Inquiry Vocabulary

Evidence
Scientific Explanation
Model
Scale Model
Prediction
Scientist
Three–Dimensional (3-D)
Two–Dimensional (2-D)

Space Science Vocabulary

Atmosphere
Satellite
Orbit
Diameter
Sphere
System

4. Predict distance between Earth and Moon. Ask them to show with their hands how far apart they think the Earth and Moon would be if they were the size of these 3-D models. (If they think the distance is greater than they can reach, tell them to point in both directions.)

5. Summarize predictions. Quickly do a visual scan of the room, and report an approximation of the range of predictions. For example, "I'm seeing everything from about 1 cm to farther than a person can reach."

6. The Moon would be 12 cm away from the Earth. Reveal the distance at this scale by holding the card with the tiny 3-D Moon about 12 cm away from the 3-D model Earth.

7. Kittinger's jump at this scale is 1/100th of 1 mm. Say, "If the Earth and Moon were this size and this far apart, Kittinger's jump would only be approximately one–one hundredth of a millimeter (less than a hair's width away from the model Earth). So even though his jump was from the greatest height any person has ever jumped down to Earth, at this scale, Kittinger's jump is too small for us to measure."

Measuring the Distance to the Sun

1. Predictions. Ask students to predict about how far it would be to the Sun at this scale, if the distance to the Moon is 12 centimeters. Accept their answers, and then tell them that the distance to the Sun at this scale would not fit into the classroom!

2. Go outside the classroom, bringing 3-D Earth, Moon, and Sun models. Have one student bring the Earth and Moon models, so you can compare the distances, and have another student bring your 44–cm diameter model of the Sun.

3. Prepare to pace off the distance to the Sun. Gather the class where you decided to begin pacing the distance to the Sun. A giant step is about 1 meter. They are going to pace out how many meters it is from the Earth to the Sun, using this model.

At this scale, 1 meter = 3,000,000 km. Because each giant step equals 3,000,000 km, 10 steps = 30,000,000 km.

4. Pace off 10 steps (30,000,000 km). Have a student hold the Earth and Moon models about 12 cm apart. Remind students that the Moon orbits the Earth, but for this model it will stay stationary. Tell the student holding the 3-D model Sun to pace off 10 giant steps away from the Earth and stop. Tell students that, at this scale, 10 giant steps represent 30,000,000 kilometers.

Unit Goals

Size: Some sky objects are relatively small, and some are huge.

Distance: Some objects are relatively close to Earth and some are very far away.

Distance of sky objects from us affects their apparent size: Large objects appear small when far away.

What One Teacher Said

"They absolutely LOVED going outside and pacing off the distance to the Sun for our scale model. We paced it from the classroom wall to the field. They were so excited they couldn't just stand on the sidelines and watch someone else pace the distance: they all spontaneously joined in. After we reached the Sun, they wondered how the distance would stack up against the length of our school building. So we paced that distance off, and it was almost the entire length of the building. I think they are going to remember this."

5. Is this the distance to the Sun? Looking at the 12 cm Earth–Moon distance for comparison, ask who thinks that 30,000,000 kilometers might be the distance to the Sun. Say that the distance to the Sun is more than this.

6. Pace off 10 more steps. Keeping the class where it is, choose a new student to join the first student, hold the Sun, and pace off another 10 meters. When the pair is 20 meters away, ask the students if they think that this distance (60,000,000 km) represents how far away the Sun is at this scale. [No.]

7. Pace off 30 more giant steps. Tell the pair to take 30 more giant steps, and then stop. When the students holding the model Sun are 50 meters away, tell them that this is how far away the Sun would be at this scale. The Sun is about 150,000,000 kilometers from Earth.

8. At this distance, the Sun looks smaller. Ask them if the Sun looks smaller at this distance. (It should look about the size of the real Sun in the sky.) Have the class return to the classroom.

Evidence Circles Discuss Relative Distances of Earth, Moon, and Sun

1. Show Sun–Earth line. Draw a horizontal line on the board about four feet long, or, on the overhead projector, display a horizontal line that is most of the width of a transparency. Make a point at each end of the line. Write “Earth” at one end, and “Sun” at the other end.

Earth .____. Sun

2. Where would the Moon be? Tell the class that this model represents distances among objects, not the sizes of objects. Ask, “If the Earth and Sun were this far apart, where would the Moon be?” Tell them that their challenge will be to make an X somewhere on the line to show where the Moon would be.

Note: If you have one or more students in your class with a reputation for knowing a lot about space science, it’s probably best that you not choose them to come forward. Otherwise, other students may simply follow their lead, rather than try to figure it out for themselves.

3. Three volunteers show where on the line they think the Moon would be and explain their reasoning. One at a time, ask three students to come up, make an X on the line to show where they think the Moon would be in the model, and explain to the class why they chose that location.

Unit Goals

Size: Some sky objects are relatively small, and some are huge.

Distance: Some objects are relatively close to Earth and some are very far away.

Distance of sky objects from us affects their apparent size: Large objects appear small when far away.

TEACHER CONSIDERATIONS

TEACHING NOTES

Noticing the Sun: If it's a sunny day, you might have the students notice the light and warmth they feel from the Sun when they are outside. Have them look at the Earth and Sun models and think about how far the sunlight traveled before reaching their skin. At the speed of light (300,000 km per second), it took about 8.5 minutes to get to Earth and touch their skin.

QUESTIONNAIRE CONNECTION

The activities in this session deal directly with question #3 on the Pre Unit 1 Questionnaire.

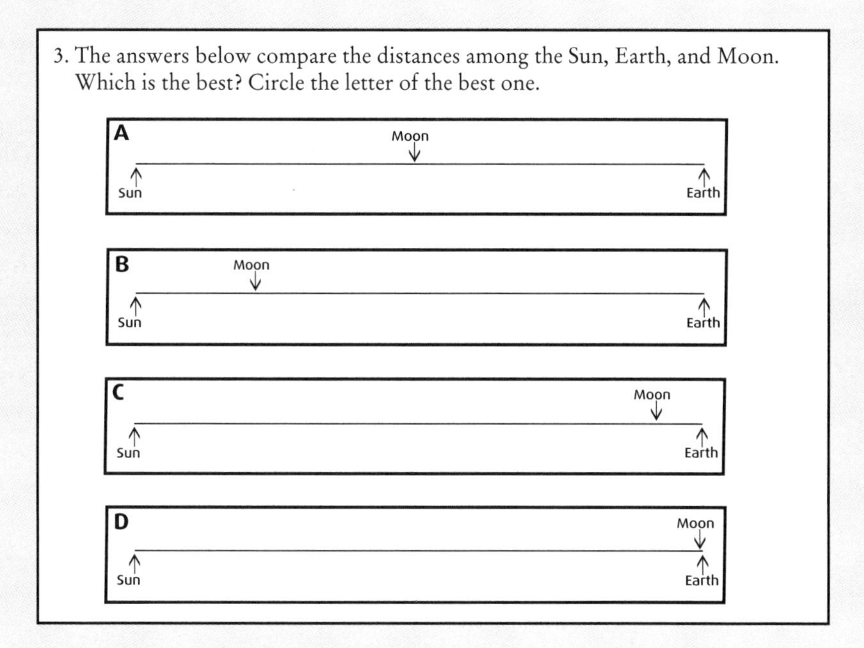
3. The answers below compare the distances among the Sun, Earth, and Moon. Which is the best? Circle the letter of the best one.

ASSESSMENT OPPORTUNITY

A Quick Check for Understanding: During the Sun–Earth line activities in this session, you may notice that students need review of the real distances involved. Or, students may grasp the relative distances, but be confused about how to represent them on the linear scale model.

PROVIDING MORE EXPERIENCE

If students have trouble with the linear scale model that represents distance, you might give them more practice using familiar distances in town. For example, you could draw a line to represent the distance from a student's home to school, and have students put an X where the school office (or any other intermediate landmark) would be.

What One Teacher Said

"The final pace off is what solidified the concept of the Sun's distance."

Key Vocabulary

Science and Inquiry Vocabulary

Evidence

Scientific Explanation

Model

Scale Model

Prediction

Scientist

Three–Dimensional (3-D)

Two–Dimensional (2-D)

Space Science Vocabulary

Atmosphere

Satellite

Orbit

Diameter

Sphere

System

EVIDENCE ABOUT DISTANCES . . .

Name:________________________

Evidence About Distances Among the Earth, Moon, and Sun

How far from Earth are the Moon and Sun?

- The Moon is about 385,000 (three hundred eighty five thousand) kilometers from Earth.
- The Sun is about 150,000,000 (one hundred fifty million) kilometers from Earth.
- The Sun is about 400 times as far from Earth as the Moon is.

How long would it take to travel to the Moon and Sun at the speed of a fast passenger jet (1,000 kilometers per hour)?

- It would take 16 days to travel to the Moon.
- It would take 6,000 days to travel to the Sun.

How far from the Earth are the Moon and Sun in the class scale model?

- If the Moon was 12 cm away from the Earth in a scale model, the Sun would be 48 meters away from the Earth.

How long did it take the Apollo astronauts to reach the Moon?

- Apollo astronauts reached the Moon in less than four days even though they coasted almost the entire distance, fighting against the gravitational pull between their spacecraft and Earth. They got a fast start!

4. Students read evidence sheet and think about where they would put the Moon in the model. Without concluding what the answer should be, pass out a copy of the sheet *Evidence About Distances Among the Earth, Moon, and Sun* to each student. Tell them to read the evidence on the sheet, and, using this evidence and the evidence from pacing the distance to the Sun, to decide where they think the Moon would go on the line.

5. Introduce evidence circle procedure. Tell the class that they will be discussing the question in small teams called *evidence circles.* As scientists, they will listen to one another, ask questions, and try to agree. The question is, "If the Earth and Sun were this far apart, where would the Moon be?" Remind them that their challenge will be to make an X on the line to show where the Moon would be in this model.

6. Explain the procedure:

a. One student says what he thinks, and the reasons why.

b. Other students who agree add their reasons.

c. Then, each student who disagrees says why, and presents her reasons.

d. The group members discuss with one another to see if they can come to agreement. If no one disagrees, they can talk about all the evidence that makes them all convinced of their view.

7. Good scientists are open to changing their minds based on evidence. The main point of evidence circles is to think about and discuss ideas and evidence in order to find the best explanation for something. Say that one of the signs of a true scientist is the ability to listen to others and change your mind when you find that what you think doesn't match the evidence.

Unit Goals

Size: Some sky objects are relatively small, and some are huge.

Distance: Some objects are relatively close to Earth and some are very far away.

Distance of sky objects from us affects their apparent size: Large objects appear small when far away.

TEACHER CONSIDERATIONS

PROVIDING MORE EXPERIENCE

Practice Procedures for Evidence Circles: The procedures for evidence circles can be a challenge the first time. Students may not be accustomed to listening and taking turns systematically. (Many adults could use practice with these skills too!) Most students also need practice in the skill of using evidence to back up their reasoning.

There will be another evidence circle in Session 1.9. If you think your students need it and if time allows, you could provide more practice with the evidence circle procedures. You could use the evidence circle procedure to discuss the example on the practice questionnaire in Session 1.1:

Compare the distances among a person's feet, knees, and head. Put an X where you think the knees would be on the line below.

head .____. feet

What Some Teachers Said

"I like the use of the Evidence Circles. It really forced my students to answer why they chose certain answers over others. The overheard discussions seemed to stay more on task, they were more polite with each other, and some were very encouraging to one another."

"The evidence circles went smoothly. They will need more practice with this because it causes them to analyze results and form an opinion based on data and work to persuade others using evidence. This is generally a difficult thing for 5th graders to do as their developmental level is just beginning to allow for this type of reasoning. But, it is very good practice for evaluating."

"Students enjoyed evidence circles. They respected each other and felt confident when sharing their views."

8. Practice the procedure. Ask, "What if a student in your evidence circle says, for example, 'I think the Moon would go here.' Is that all he needs to say?" [No, he needs to give a reason.]

9. Circulate during discussions. Go around to different teams to make sure that they are following the procedure and using evidence to back up their thinking.

Large-Group Discussion of Where to Put the Moon in the Model

1. Each group shares their answer and reasoning with the class. When evidence circles have finished their discussions, get the attention of the class. Ask a team that came to agreement to send someone forward to show where they would put the Moon on the line and share their thinking and the evidence for it. Ask other teams to do the same, each time erasing the X of the previous group.

2. It's hard to transfer from one model to another. When several teams have presented, tell them that it can be very difficult to transfer information from one model (such as the 3–D model) to another (such as the X on the line).

3. The X should be as close to Earth as possible. Remind them that although the Moon is very far from us, the Sun is so much farther from the Earth that *in comparison,* the Moon is not far at all. Draw an X as close as you possibly can to the point labeled "Earth," almost on top of it. Tell them that this is how far away the Moon would be from the Earth in this model.

4. Add to concept wall. Help the students reflect on what they have learned from the activities in this session. Add the following key concepts to the concept wall:

The atmosphere is thin compared to the size of the Earth—like the skin of an apple.
The Sun is much farther from Earth than the Moon is.

Unit Goals

Size: Some sky objects are relatively small, and some are huge.

Distance: Some objects are relatively close to Earth and some are very far away.

Distance of sky objects from us affects their apparent size: Large objects appear small when far away.

TEACHER CONSIDERATIONS

OPTIONAL PROMPTS FOR WRITING AND DISCUSSION

You may want to have students use the prompts below for science journal writing during class or as homework. Or they could be used for a discussion or during a final student sharing circle.

- Tell one thing that you learned about the Earth, Moon, and Sun today.
- Describe how it would look, feel, and be if you could travel in a balloon very high in the sky, and then skydive back down to Earth.

TEACHING NOTES

For quick reference, here are the scale model measurements of sizes and distances of objects examined in this session.

	Actual	Scale Model
Sun Diameter	~1,400,000 km	44 cm
Earth Diameter	~13,000 km	4 mm
Moon Diameter	~3,000 km	1 mm
Earth–Moon Distance	~385,000 km	12 cm
Earth–Sun Distance	~150,000,000 km	50 m
Earth–Satellite #2 Distance	~30,000 km	1 cm
Height of Atmosphere	~500 km	1/6th of 1 mm
Highest Skydive	~31 km	1/100th of 1 mm

Key Vocabulary

Science and Inquiry Vocabulary

Evidence

Scientific Explanation

Model

Scale Model

Prediction

Scientist

Three–Dimensional (3-D)

Two–Dimensional (2-D)

Space Science Vocabulary

Atmosphere

Satellite

Orbit

Diameter

Sphere

System

SESSION 1.9

How Our Scale Ideas Have Changed

Overview

Looking at an image of Jupiter, students discuss how the image can be so large. They learn that either through magnification or through getting closer with a spacecraft, or a combination of the two, we can see larger images of sky objects. They learn that telescopes are very useful tools for space scientists.

The next activity is a demonstration designed to review relative sizes of sky objects and solidify students' understanding of how apparent size depends on distance from the viewer. The teacher holds up a dot representing the size of Venus, and students compare it with a "Planet X" dot. The teacher backs up, and the students note when the Venus dot is far enough away that it appears to be the same size as Planet X. They discuss how far away the models representing Jupiter and the Sun would need to be to appear as small as they do in the sky. Students also review that although our Sun is a medium-sized star, it looks much bigger than other stars because it is much closer to us.

In evidence circles of four, the students are given three inaccurate models of the Solar System. They apply the evidence that they have gathered about scale as they evaluate each model for accuracy in terms of size and distance. Their discussion is continued with the whole class. Finally, to conclude the unit, the students take the *Post–Unit 1 Questionnaire* on size and distance to find out how their ideas have changed.

How Our Scale Ideas Have Changed	Estimated Time
Introducing telescopes	5 minutes
Comparing the sizes and distances of planets and stars	10 minutes
Evidence circles	30 minutes
Taking the *Post–Unit 1 Questionnaire*	15 minutes
TOTAL	**60 minutes**

What You Need

For the class

- ❑ overhead projector or computer with large screen monitor/LCD projector
- ❑ transparency from the transparency packet or CD–ROM image of Jupiter

Unit Goals

Size: Some sky objects are relatively small, and some are huge.

Distance: Some objects are relatively close to Earth and some are very far away.

Distance of sky objects from us affects their apparent size: Large objects appear small when far away.

SAMPLE CLASSROOM CONCEPT WALLS

Key Vocabulary

Science and Inquiry Vocabulary

Evidence

Scientific Explanation

Model

Scale Model

Prediction

Scientist

Three–Dimensional (3-D)

Two–Dimensional (2-D)

Space Science Vocabulary

Atmosphere

Satellite

Orbit

Diameter

Sphere

System

- ❑ 1 dot sheet with thirty–two 1–mm Planet X dots, one 4–mm Venus dot, and one 4.8–cm Jupiter dot, from the student sheet packet
- ❑ 2–D paper model of Sun used in Sessions 1.4 and 1.8
- ❑ sentence strips for 3 key concepts
- ❑ wide-tip felt pen

For each group of four students

- ❑ 1 *Size and Distance of Solar System Objects* sheet, from the student sheet packet

For each student

- ❑ 1 metric ruler
- ❑ 1–mm Planet X dot from dot sheet from the student sheet packet
- ❑ 1 copy of the *Post–Unit 1 Questionnaire* from the student sheet packet
- ❑ 1 *Inaccurate Models* student sheet from the student sheet packet

Getting Ready

1. Arrange for the appropriate projector format (computer with large screen monitor, LCD projector, or overhead projector) to display images to the class.

2. If you will not be using the CD-ROM, make an overhead transparency of the image of Jupiter.

3. Make a copy of the *Inaccurate Models* sheet for each student.

Optional: In addition to the three drawings of the Solar System on the *Inaccurate Models* sheet, you might want to collect other Solar System drawings or 3-D Solar System models with inaccuracies regarding size and distance. It is usually not difficult to locate examples of models with inaccuracies in books or on posters. For example, many models portray the planets as very close together or show the Sun as smaller than it is, etc.

4. Make a copy of the *Size and Distance of Solar System Objects* sheet and cut in thirds so that you have one chart for each group of four students.

5. Make 1 copy of the dot sheet, and cut it up so that you have one Planet X dot per student. Cut out the Venus dot and the Jupiter dot, and have them handy for your demonstration.

6. Set aside one metric ruler per student.

Unit Goals

Size: Some sky objects are relatively small, and some are huge.

Distance: Some objects are relatively close to Earth and some are very far away.

Distance of sky objects from us affects their apparent size: Large objects appear small when far away.

THE KEY CONCEPT WALL

BY THE END OF UNIT 1, YOUR CONCEPT WALL SHOULD LOOK SOMETHING LIKE THIS:

WHAT WE HAVE LEARNED ABOUT EVIDENCE AND MODELS

1. Evidence is information, such as measurements or observations, that is used to help explain things.

2. Scientists base their explanations on evidence.

3. Scientists question, discuss, and check each other's evidence and explanations.

4. Scientists use models to help understand and explain how things work.

5. Space scientists use models to study things that are very big or far away.

6. Models help us make and test predictions.

7. Every model is inaccurate in some way.

8. Models can be 3-dimensional or 2-dimensional.

9. A model can be an explanation in your mind.

WHAT WE HAVE LEARNED ABOUT SPACE SCIENCE

1.1

Earth is surrounded by an atmosphere of air.

Beyond Earth's atmosphere is what we call space.

People have been wondering and learning about space for a long time.

1.4

Some objects in the sky, such as the Sun, Moon, stars and planets, are very large.

Other objects in the sky, such as birds, satellites and airplanes, are relatively small.

The Earth is very large.

The Moon is very large, but not as large as the Earth.

The Sun is super huge compared to the Earth.

The Sun is a star. Compared to other stars it is medium sized.

1.5

How big something looks and how big it really is can be very different.

An object looks bigger when it's closer. An object looks smaller when it's farther away.

The Sun looks bigger than other stars because it's a whole lot closer.

The Sun looks the same size as the Moon because it's much farther away than the Moon.

1.6

There are many things in the Universe that are much larger than the Sun.

1.7

Some objects we see in the sky, such as birds and airplanes, are relatively close.

The Moon is very far from us compared to objects in Earth's atmosphere.

1.8

The atmosphere is thin compared to the size of the Earth – like the skin of an apple.

The Sun is much farther from Earth than the Moon.

1.9

Other stars are much farther away from Earth than the Sun is.

Most things in the Universe are much farther away from Earth than the Sun is.

Telescopes can be used to make objects appear brighter and larger.

7. Write the following key concepts on sentence strips to post during the session:

Other stars are much farther away from Earth than the Sun is.
Most things in the Universe are much farther away from Earth than the Sun is.
Telescopes can be used to make objects appear brighter and larger.

Introducing Telescopes

1. Show image of Jupiter. Project the image of Jupiter on a transparency or using the CD-ROM. Ask, "Is this how Jupiter looks in the sky?" [No, from Earth and with the unaided eye, Jupiter looks like a bright star.]

2. How do we get images such as this? Ask, "How do you think we got this close-up picture of Jupiter?" [This image came from NASA's *Cassini* spacecraft on December 7, 2000. It could have come from a spacecraft that went near Jupiter, or from a telescope on Earth, or from a space-based telescope.]

3. Telescopes make things look brighter and bigger. Tell them that *telescopes* have lenses that make things that are very far away look brighter and bigger. Telescopes are very useful tools for scientists to study things in space. The invention of the telescope (in the early 1600s) was a huge historic advance in the science of astronomy!

Reviewing How Distance Affects Apparent Size

1. Two ways objects can appear larger. Reiterate that an object can appear larger in two ways: we can magnify it with a telescope or we can go closer to the object (or both).

2. Review why the Moon and Sun appear to be the same size. Remind students that their measurements of models in Session 1.4 showed that the Sun is really about 400 times bigger across than the Moon. Review, "Why do the Moon and Sun appear to be almost the same size when seen from Earth?" [The Moon is much closer to us than the Sun is.]

Unit Goals

Size: Some sky objects are relatively small, and some are huge.

Distance: Some objects are relatively close to Earth and some are very far away.

Distance of sky objects from us affects their apparent size: Large objects appear small when far away.

TEACHING NOTES

Other Images of Planets and Telescopes: If you have time, you may want to show students some of the other beautiful NASA images of space objects that can also be seen with the naked eye—especially Mars, Saturn, and Venus. You may also want to seek out and show pictures of the telescopes and spacecraft that obtained these exciting images. Please see the *Resources* section, pages 457-459, for information on how to get these images from NASA.

More on Telescopes: Understanding how telescopes work is beyond the scope of this unit. Later, in middle school, the GEMS guide *More Than Magnifiers* would be a great way for students to learn more about telescopes and magnification. In that unit, students construct and use simple telescopes (as well as cameras and projectors) using two lenses. They also learn important concepts about light and optics. For more information on telescopes, see page 455 in the *Background* section.

Session 1.9 Transparency

NASA *Cassini* spacecraft, December 7, 2000
Jupiter's moon Europa is casting the shadow on the planet.

Jupiter

3. The Sun is 400 times farther away than the Moon. Tell them that although the Sun is about *400 times* bigger across than the Moon, it also happens to be about four hundred times farther away, which makes it look as though it's about the same size as the Moon when seen from Earth.

Comparing the Sizes and Distances of Planets and Stars

1. A demonstration to compare distances of planets and stars. Tell students that they're going to do an activity now to compare the distances of some planets and stars.

2. Hand out Planet X dots. Give each student a 1–mm Planet X dot. Say that the dot is a 2–D scale model of a make–believe Planet X—not a real planet. Tell them that the model Planet X is 1 mm in diameter.

3. Students bring Planet X and stand in groups at end of room. Tell them to bring their Planet X dots with them and stand in a group at one end or corner of the room. Have them hold their Planet X at arm's length, and say that what they are seeing represents the view of this "planet" from Earth.

4. Compare Venus dot with Planet X dot. Stand close to the students, and hold up the Venus dot so that your students can see it. Say that the dot is a scale model of Venus—a real planet. Point out that this dot is about four times as large as the dot representing Planet X. Briefly hold a Planet X dot next to the Venus dot for comparison.

5. Tell them that you are going to move Venus away from them. Explain that when Venus appears to be the same size as Planet X, they should say, "Now."

6. Slowly move Venus away from students. Holding the Venus model, slowly move backward, away from the students. Because they are all standing in different positions, the position at which the two planets appear the same size will not be the same for everybody. If necessary, keep moving till you reach a wall. Tell them to raise their hands if this was not enough distance to make Venus appear the same size as Planet X. If any hands are raised, ask how much farther they think you'd need to move away for them to appear the same size.

Unit Goals

Size: Some sky objects are relatively small, and some are huge.

Distance: Some objects are relatively close to Earth and some are very far away.

Distance of sky objects from us affects their apparent size: Large objects appear small when far away.

TEACHER CONSIDERATIONS

TEACHING NOTES

Adjusting for size and space: Students should be standing where they can hold their dot at arm's length and also see you. They can be in a line or a cluster.

DOTS

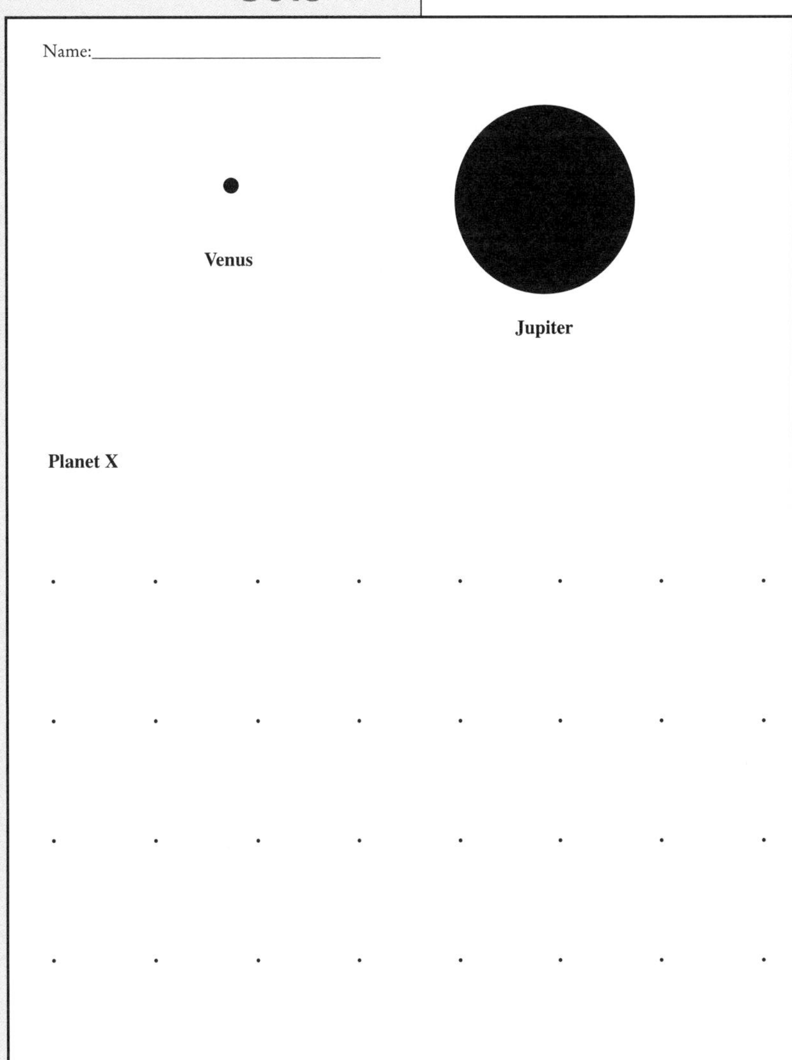

QUESTIONNAIRE CONNECTION

The *Comparing the Sizes and Distances of Planets and Stars* activity reviews questions #2 and #4 on the *Pre-Unit 1 Questionnaire.*

2. Why does the Sun look much bigger than the stars we see at night? Circle the letter of the best answer.
   A. The Sun is much bigger than the stars.
   B. The Sun looks bigger because it is closer to us than the stars.
   C. The Sun looks bigger because it is farther away from us than the stars.
   D. The Sun can be seen only in the daytime.

4. Why do the Sun and Moon look as though they are about the same size in the sky? In your answer, explain how big and how far away they are.

7. Jupiter looks smaller than Venus in the sky. Ask if anyone remembers the real size of Jupiter from Session 1.6. [It is much bigger than Venus, but much smaller than the Sun.] Now, hold up the Jupiter model, which is about 12 times larger in diameter than Venus, but looks smaller than Venus in the sky. Ask students to imagine how far away Jupiter would have to be to appear smaller than Venus and Planet X.

8. Jupiter would have to be at least 120 meters away to appear a little smaller than Venus. Tell them that although Jupiter is almost 12 times as large as Venus, it actually looks a little smaller than Venus when seen in the sky.

9. Hold up the Sun model. Now, hold up the 44–cm 2–D model of the Sun used in Sessions 1.4 and 1.8. Ask them to remember how far away this model Sun was from Earth in this model (50 meters, as in the schoolyard model in Session 1.8) and ask them to remember how big this model of the Sun looked at that distance away. [It looked about the same size as the real Sun looks in the sky.]

10. Have everyone sit down. Collect the Planet X dots.

11. The Sun is a star. Other stars look tiny because they are so far away. Tell them that even though the Sun is incredibly far away from Earth, it is still our closest star. Our Sun is medium-sized for a star, but all other stars look tiny in comparison, because they are so much farther away. Imagine how far away the 44–cm model Sun would have to be to look like a dot-sized star!

After the Sun, the next closest star, Proxima Centauri, is about 42,000,000,000,000 (forty-two trillion) kilometers from Earth. In the current scale model, that would be 14,000 kilometers away (more than the diameter of Earth!), and other stars are much, much farther away.

12. Without other evidence, people looking at the sky might think that the Moon is huge and Jupiter is tiny. Tell them that a person just looking at the sky might think that the Moon is huge and that Jupiter is tiny. And just from observing Venus in the sky, people might think that it's larger than stars, but, of course, it's much smaller. With evidence that scientists have gathered about the real sizes, we can now understand that Venus is much smaller than stars, but it *looks* bigger.

Unit Goals

Size: Some sky objects are relatively small, and some are huge.

Distance: Some objects are relatively close to Earth and some are very far away.

Distance of sky objects from us affects their apparent size: Large objects appear small when far away.

TEACHER CONSIDERATIONS

TEACHING NOTES

Observing Venus in the Sky: If you know that the planet Venus will be visible the night after you present this session, ask your students if they think that Venus will be larger or smaller than the stars. Larger or smaller than the Moon? Tell them where to find Venus in the sky, and assign them to observe and see for themselves that night.

Key Vocabulary

Science and Inquiry Vocabulary

Evidence

Scientific Explanation

Model

Scale Model

Prediction

Scientist

Three–Dimensional (3-D)

Two–Dimensional (2-D)

Space Science Vocabulary

Atmosphere

Satellite

Orbit

Diameter

Sphere

System

13. Post key concepts. Add the final Unit 1 key concepts to the concept wall:

Other stars are much farther away from Earth than the Sun is.
Most things in the Universe are much farther away from Earth than the Sun is.
Telescopes can be used to make objects appear brighter and larger.

Evidence Circles: Evaluating Models

1. Show *Inaccurate Models* sheet. Hold up the *Inaccurate Models* sheet. If you have gathered other pictures of the Solar System or 3-D models, show these, too. Say that sometimes models show inaccurate sizes and distances of space objects.

2. Evidence circles evaluate the accuracy of these models in terms of size and distance. Tell them that they will work in evidence circles. Their team will pick one of the models and discuss what is inaccurate about the model in terms of both *size* and *distance*. Review the procedure for evidence circles:

a. Pick one model to discuss. (Model A, B, or C on the Inaccurate Models sheet, or another model that you have provided.)

b. Each student gets a turn to say one way the model is inaccurate (not exactly the same as the real thing) and why.

c. The group discusses the evidence with one another to see if they can agree.

d. Each student on the team writes down all the ways the model is inaccurate.

3. Focus on size and distance. Say that it is fine to discuss anything they find that is wrong about the model. However, they should be sure to discuss what is inaccurate about the sizes and distances of objects in the picture.

Unit Goals

Size: Some sky objects are relatively small, and some are huge.

Distance: Some objects are relatively close to Earth and some are very far away.

Distance of sky objects from us affects their apparent size: Large objects appear small when far away.

TEACHER CONSIDERATIONS

TEACHING NOTES

Why use the Solar System diagrams for the inaccurate models activity? The relative sizes of objects in the Solar System were introduced in Session 1.6, but the focus of Unit 1 and the rest of the units in this grade 3–5 *Sequence* is not on the Solar System, but on the Earth, Moon, and Sun system. Nevertheless, it works well to use models of the Solar System in this closing activity. Solar System models are very commonly seen in classrooms and textbooks, and they are almost always full of inaccuracies, especially about the sizes and distances of the objects pictured. Students enjoy finding inaccuracies, and have plenty of evidence from Unit 1 on which to base their opinions—especially about inaccuracies in size and distance.

***Please note:* In the *Space Science Sequence* for grades 6–8,** students do study the Solar System in depth. In that unit, students will again have an opportunity to examine a common model of the Solar System for inaccuracies.

INACCURATE MODELS

Name:____________________________

INACCURATE MODELS

A

B

Neptune
Pluto
Uranus
Saturn
Jupiter
Mars
Earth
Venus
Mercury

C

Uranus
Asteroids
Neptune
Earth
Venus
Saturn
SUN
Mercury
Mars
Jupiter
Pluto

Student Sheet—Space Science Sequence 1.9

If needed, review models: If necessary before the evidence circle activity, refer to the key concepts about models on the concept wall, and review the idea that all models are inaccurate in some way, or they would be the real thing. Remind students that *inaccurate* means that the model is not exactly the same as the real thing.

SIZE AND DISTANCE OF SOLAR SYSTEM OBJECTS

Size and Distance of Solar System Objects

	Diameter	Distance from the Sun
Sun	1,391,000 km	0 km
Mercury	4,879 km	57,909,175 km
Venus	12,104 km	108,208,930 km
Earth	12,756 km	149,597,890 km
Mars	6,794 km	227,936,640 km
Jupiter	142,984 km	778,412,020 km
Saturn	120,536 km	1,426,725,400 km
Uranus	51,118 km	2,870,972,200 km
Neptune	49,528 km	4,498,252,900 km
Pluto	2,302 km	5,906,380,000 km

4. Use *Size and Distance of Solar System Objects* sheet for evidence. Say that they have gathered evidence during the unit about the real sizes of these objects, especially the Sun, Earth, and Moon. Tell them that you will give each team a *Size and Distance of Solar System Objects* sheet with more evidence that they can use to help them decide if the models are inaccurate.

5. Evaluate other models. Say that if their evidence circle finishes earlier than others, they can go on to discuss one (or both) of the other models on the sheet. Pass out one *Inaccurate Models* sheet to each student, and one *Size and Distance of Solar System Objects* sheet to each team.

6. Circulate as they work.

7. Large–group discussion of accuracy of the models. When most teams have finished, lead a large–group discussion. Go through the models one at a time, asking students to share inaccuracies of size and distance in each model.

Taking the *Post–Unit 1 Questionnaire*

1. Their ideas may have changed. Point to the concept wall, and acknowledge all the work that students have done as scientists and all that they have learned during the past nine class sessions. Remind them that the ability to change ideas based on evidence is a sign of a good scientist. The students have gathered and discussed lots of evidence about the sizes and distances of sky objects, and they may have changed some of their ideas. Say that they now get to take the Unit 1 Questionnaire again to see how their ideas may have changed.

2. Same questions, different order. Explain that the questions are exactly the same as on the *Pre–Unit 1 Questionnaire*, except the order of the questions is different. And within a question, some of the answers they can pick from have been reordered, too. Explain that this is so that they will pay attention to what the question is asking, rather than just remembering to circle the same letter as before.

TEACHER CONSIDERATIONS

OPTIONAL WRITING ACTIVITY

This writing assignment could be used before or after the evidence circle activity:

Choose Model A, B, or C. Write down what is inaccurate about the sizes and distances of objects in the model. Write down the evidence that shows that it is inaccurate.

1. Preparation for evidence circles: If used before the activity, it will give students a chance to think for themselves about what is inaccurate in the model and jot down the evidence, before sharing with their team. This may make the evidence circle discussion richer, especially for students who are not experienced in making evidence-based explanations.

2. Embedded Assessment: After the evidence circle activity, this writing assignment could be used as an assessment of their skill in making evidence-based explanations and scored using the general rubrics provided on page 66. It could also give you some insight into student understanding of the relative sizes and distances of objects in the Solar System.

QUESTIONNAIRE CONNECTION

A scoring guide is included on page 71 for the *Post-Unit 1 Questionnaire.*

OPTIONAL PROMPTS FOR WRITING OR DISCUSSION

You may want to have students use the prompts below for science journal writing during class or as homework. Or they could be used for a discussion or during a final student sharing circle.

- Describe what you think are the most important things you've learned during the whole unit on size and distance of objects in space.
- Why do you think it is so difficult to make an accurate model of the Solar System?

What Some Teachers Said

"The students have a good understanding of evidence circles. They worked well with others and they did a wonderful job explaining their pieces of evidence."

"The inaccurate models were a hit."

Key Vocabulary

Science and Inquiry Vocabulary

Evidence

Scientific Explanation

Model

Scale Model

Prediction

Scientist

Three–Dimensional (3-D)

Two–Dimensional (2-D)

Space Science Vocabulary

Atmosphere

Satellite

Orbit

Diameter

Sphere

System

POST–UNIT 1 QUESTIONNAIRE, PAGE 1

Name:______________________________

Post–unit 1 Questionnaire

1. Why do the Sun and Moon look as though they are about the same size in the sky? In your answer, explain how big and how far away they are.

____

____

____

____

2. The answers below compare the distances among the Sun, Earth, and Moon. Which is the best? Circle the letter of the best one.

A. Moon / Sun / Earth

B. Moon / Sun / Earth

C. Moon / Sun / Earth

D. Moon / Sun / Earth

3. Why does the Sun look much bigger than the stars we see at night? Circle the letter of the best answer.
A. The Sun looks bigger because it is closer to us than the stars.
B. The Sun is much bigger than the stars.
C. The Sun looks bigger because it is farther away from us than the stars.
D. The Sun can be seen only in the daytime.

Continued on next page

3. Review the questionnaire guidelines: Don't help one another. The questionnaire is designed to find out how *each* student's ideas have changed. Tell them that if they don't know all the answers, that's fine. They should just think about what they have learned and do their best.

4. Distribute pencils and questionnaires.

5. Discuss their answers. When everyone has finished, collect the questionnaires. Ask students to share their answers and reflect on what they have learned during Unit 1. Congratulate them on how much they have learned about the sizes and distances of objects in the sky.

POST-UNIT 1 QUESTIONNAIRE, PAGE 2

Name:________________________________

Post-unit 1 Questionnaire continued

4. One of the pictures below shows the correct sizes of the Sun, Earth, and Moon compared with one another. This question is about the real sizes, *not* how big they look in the sky. Which is the best? Circle the letter of the best one.

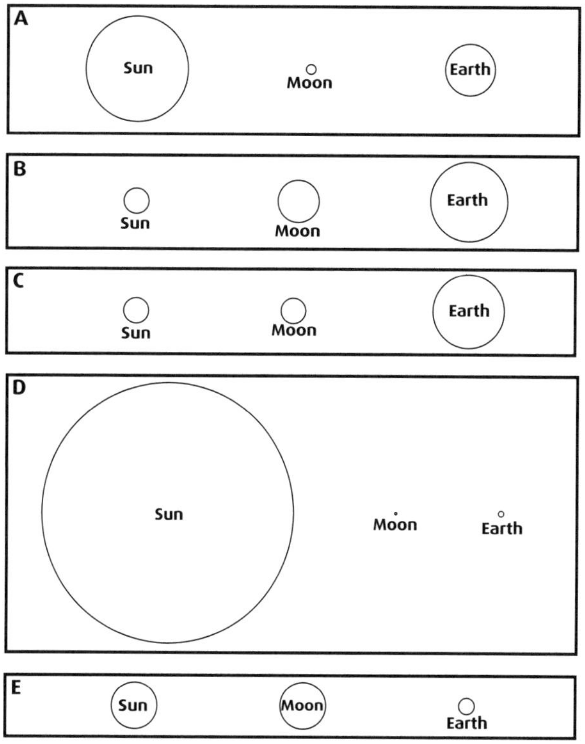

A — Sun, Moon, Earth

B — Sun, Moon, Earth

C — Sun, Moon, Earth

D — Sun, Moon, Earth

E — Sun, Moon, Earth